Coiled Tubing Operations at a Glance

What Do You Know about Coiled Tubing Operations?

KHOSROW M. HADIPOUR

To order additional copies of this book, contact:
Xlibris
1-888-795-4274
www.Xlibris.com
Orders@Xlibris.com

ISBN:	Softcover	978-1-7960-7186-3
	Hardcover	978-1-7960-7187-0
	EBook	978-1-7960-7185-6

Library of Congress Control Number: 2019918683

Print information available on the last page

Rev. date: 01/10/2020

Professional Petroleum Engineer

The author's presentation on the subject material is based on forty-two years of offshore and onshore downhole experience in drilling, completion, production, fracturing, downhole fishing, sidetrack drilling, remedial cementing, coiled tubing intervention operations, oil and gas remedial workover repairs, artificial fluid lift, gravel packing, plug and abandonment, and consulting experience while working for Gulf Oil company, Chevron USA, Pennzoil, Devon Energy, and AmeriCo Energy resources in Texas, Mississippi, Louisiana, New Mexico, the Gulf of Mexico, and Venezuela.

The written material is presented as basic information about coiled tubing only and is not intended to be a work procedure or a guideline for anyone to follow. We are not responsible for the information herein.

Kendall Lease
Well # 1
Durkee Field

Application of Coiled Tubing in Oil and Gas Well Interventions

Coiled tubing is aggressive and fast growing in most wellbore intervention solutions. Maximizing well performance and oil and gas production are the coiled tubing achievements. Today's advanced coiled tubing operation offers several unique, efficient, and cost-effective well intervention applications in oil and gas across the world.

Coiled tubing offers safety, well-controlled measures, and cost-saving alternatives in the oil and gas wellbore operations listed below:

- Oil and gas wellbore cleaning and pressure control techniques
- Snubbing and stripping operations (coiled tubing is the next best solution)
- Cost-effective in horizontal drilling
- Completion and through tubing gravel packing and fracturing
- Oil and gas stimulation and through tubing fracturing
- Fishing and remedial applications
- Cementing and borehole cleaning
- Zone isolation
- Hard-scale milling and junk snatcher techniques

You name it, coiled tubing can do it!

You will read about the advantages and limitations of coiled tubing applications as you go through this book. Oil and gas remedial interventions can be achieved using coiled tubing safely and cost-effectively without pulling the production equipment or killing a live well.

One of the major reasons for the development of coiled tubing is to prevent a time-consuming conventional workover rig from killing a well, releasing the mechanical isolation equipment, pulling the tubing string to conduct a minor remedial repair to a well, and placing the well back on production several days later.

Using standard offshore and/or onshore rigs often becomes expensive to work on a well. Some of the well interventions can be achieved in one day using coiled tubing without pulling production equipment and/or killing the well to save cost and production downtime (excellent invention and intervention).

Many challenging problems can develop in an oil and gas well as time passes. The reduction of oil and gas production because of sand and solids through perforations, saltwater disposal problems, casing problems, and through tubing gravel-packing operations are some of the few impressive intervention solutions offered by the coiled tubing industry worldwide.

What Is Coiled Tubing?

Coiled tubing is described as a long, continuous, flexible (enough) steel pipe that spools or is wound onto a large steel reel for transportation (deployed for well interventions).

The length of continuous coiled tubing spooled onto a steel reel depends on the well depth and the pipe's outside diameter. In the old days, smaller-diameter coiled tubing strings of ¾″ and 1″ were spooled and applied in wellbore cleanup operations.

The improvement and rapid expansion of coiled tubing quality operations around the world enable the transporting of larger sizes of tubing strings for deeper and higher-pressure wellbores:

- 1″, 1 ¼″, 2, 2 ⅜″, 2 ⅞″, and even 3 ½″ pipe is used for offshore and onshore intervention operations (unbelievable!)
- Smaller coiled tubing strings generally are used in shallow well depths (3,000′ to 10,000′) in oil, gas, or saltwater disposable wells with low reservoir pressure safely
- Larger-diameter coiled tubing of 1 ½″, 2 ⅜″, 2 ⅞″, or 3 ½″ is normally spooled on huge steel reels for transportation, storage, or intervention operations in higher-pressure and deeper wells offshore or onshore (major projects with higher anticipated wellbore pressure)

The length of the continuous coiled tubing must meet or exceed the depth of a well before the equipment is transported to the well location (you do not want to be short of the pipe when reaching the target depth of 12,000′).

***

The basic invention of continuous flexible tubing started in 1944 by British engineers to supply and transport fuel across water channels to the Allied armies during the war. The project was called PLUTO (pipeline under the ocean). The coiled tubing was later improved in Canada and was called Flex-Tube.

In 1970, Brown Oil Tools, Otis Engineering, and Bowen Fishing started implementing the development of coiled tubing in oil and gas operations in Texas. In 1977, I used a coiled tubing unit from Otis Engineering in Bryan–College Station, Texas, for the first time, and I was anxious and wondered what coiled tubing could do at that time.

Since 1977, coiled tubing tools and equipment have significantly improved along with trained and knowledgeable operators of today. Many engineering thoughts and studies have been implemented in the improvement of coiled tubing tools and equipment of today (coiled tubing equipment may still need improvement!). I am impressed with coiled tubing's cost-saving interventions in oil, gas, saltwater disposal, and other wellbore applications (I am still using coiled tubing today, in 2019).

Coiled tubing tools and equipment are designed to conduct specific challenging operations in oil and gas wells in safer, faster, and more cost-effective and efficient ways:

- Well control (keeping the well pressure under control while snubbing and stripping)
- Oil and gas well horizontal drilling
- Milling objects and continuously washing and circulating solids
- Washing the well and displacing liquid from wellbores (de-liquid gas wells)
- Cementing and spotting cement plugs
- Acid stimulation and reservoir fracturing
- Plugging wellbores
- Fishing and other well interventions
- Jetting and wellbore open-hole cleaning

What Is the Coiled Tubing Unit?

If you have not seen or worked on coiled tubing, you may run into or see units on the highway and/ or at truck stops.

REDZONE

SIZE
SIZE
REDZONE

The term *coiled tubing unit* refers to the whole coiled tubing packaged assembly, consisting of the service reel with spooled coiled tubing and several other components—such as the injector, the control console, the well pressure control stack, and the power supply—that are built or mounted onto a mobile trailer truck bed or skid units ready to transport to a job site.

Land-operated coiled tubing units with smaller coiled tubing strings (1″, 1 ¼″, 1 ½″, and 2″) are normally built on long mobile truck trailers. These mobile units are quick to transport and time-saving in rig-up and rig-down bases.

Almost all the components of a coiled tubing unit are built and mounted on one long single-mobile truck trailer (the coiled tubing reel with the pipe is the heaviest of all the components). The offshore coiled tubing is built on short, lighter skid units because of limited space on the offshore platforms and the crane-lifting capacities.

The application of coiled tubing in wellbore interventions in offshore operations will save you time and money. When it comes to work, a coiled tubing company will rig up on any well in different configurations.

Transporting to jack-up rigs or post-barge rigs is generally a high-cost and time-consuming operation. The coiled tubing components may be transported out using large barges with heavy-duty cranes to lift and spot coiled tubing components up on a platform.

The coiled tubing operation may be performed from a large boat, a rig barge, onto an offshore platform. Multiple barges may be utilized to transport a coiled tubing unit with related tools and supporting equipment to an offshore well site of 4–10′ of deep water.

Transporting a coiled tubing unit onto larger barges is quicker and can be carried out using large barges and tugboats similar to a land operation (if it is done safely and correctly). Use shallow draft tugboats on shallow water depths. Some tugboats draw too much water (dredging to get to the well may be needed). Tide up or down during cold-blowing winter wind can deplete most of the bay water.

***

Do You Know How Coiled Tubing Is Made?

The physical and chemical composition of coiled tubing may be different from the standard oil field tubing joints. An oil field pipe is defined as a closed cylindrical conductor that is made from steel or other elements.

In oil field operations, the steel pipe is often called tubing or pipe (we have our own slang in the oil field). The manufacture of steel tubing is derived from the raw material called "pig iron" that goes through a lengthy process to become steel. Pig iron is the basic raw material for steel making (steel sheets, I-beams, and pipes).

Several elements (alloy) will be added to the steel to improve the physical and chemical composition of the steel pipe:

- Carbon
- Silicon
- Manganese
- Phosphorous
- Sulfur
- Aluminum
- Others

(The relation between carbon and steel is like salt and water.)

A coiled tubing string is manufactured from a low-carbon alloy steel sheet, referred to as a "slit" (a long thin flat metal sheet). The tube is formed from a continuous flat strip of a 3,500′-long steel sheet, referred to as the "slit" or "skelp," to start with. During a unique manufacture pipe-making process, the precise width of a slit sheet is measured, checked, and leveled off for tubing making.

The two major methods of forming the flat sheet to the tubing shape is either butt welding and/or "bias" welding. There are basically three manufacturing techniques of seamed tubing making:

- FW (furnace welding)
- ERW (electric resistance welding)
- SAW (submerged arc welding)

Most coiled tubing strings are bias welded (a bias weld is stronger than a butt weld). Coiled tubing is a seam type. All the welded tubing strings are referred to as ERW pipes (tubing).

The steel sheet will pass through a series of forming rollers and dies that transform the strip from a flat steel sheet to round tube-shaped sections. The edges of the strip are contoured, and the pipe gap will be compressed. The longitudinal seam will be welded by high-frequency induction welding from end to end (welding by resistance to flow) as the tube passes onto the assembly line.

As the pipe passes down the assembly line with a constant speed, the tube is welded using high-electronic frequency. The coiled tube is an electric-welded tube that is manufactured in a continuous process with one longitudinal seam from end to end (the welded seam may be present from the inside of the tube).

After the tube is welded longitudinally, it passes through heater furnaces, where it is heat treated with a high temperature of 1,650 degrees or as required. The pipe then will be cooled off, straightened visually, and inspected electromagnetically for any defects.

Coiled tubing is manufactured in a wide range of grade sizes: 70 grade, 80 grade, 90 grade, 100 grade, 110 grade, and 120 grade. The most popular coiled tubing string sizes in the oil and gas business may range from 1″ to 1 ¼″, 1 ½″, 1 ¾″, 2″, 2 ⅜″, 2 ⅞″, and 3 ½″ with a wide range of thickness and weight. The continuous flexible coiled tubing string will undergo several necessary heat treatments to develop certain mechanical properties.

COILED TUBING CAPACITY

PIPE SIZE OD	WALL THICKNESS	WT/FT	BBL/FOOT
1"	0.087	0.849	0.00069
1"	0.095	0.918	0.00061
1"	0.102	0.978	0.0006
1"	0.125	1.167	0.0005
1 ¼"	0.080	0.9997	0.0012
1 ¼"	0.095	1.172	0.0012
1 ¼"	0.102	1.251	0.0011
1 ¼"	0.116	1.405	0.0010
1 ¼"	0.125	1.502	0.0010
1 ¼"	0.175	2.0082	0.0008
1½"	0.102	1.5233	0.0017
1½"	0.109	1.6234	0.0016
1½"	0.125	1.8355	0.0015
1½"	0.156	2.2391	0.0014
1½"	0.175	2.4764	0.0013
1¾"	0.102	1.7952	0.0023
1 ¾"	0.095	1.6781	0.0024
1 ¾"	0.145	2.4861	0.0021
1 ¾	0.188	3.1363	0.0018
1 ¾	0.203	3.3542	0.0017
2.0"	0.116	2.3341	0.0031
2.0"	0.125	2.5041	0.0030
2.0"	0.134	2.6731	0.0029

The pipe will be tested according to API specifications, including but not limited to the following properties:

- Yield strength
  - Deviation from stress to strain
  - Pipe strength under load before permanent deformation
- Tensile strength
  - Maximum load until pipe ruptures
- Pipe hardness
  - Resistance to plastic
  - Resistance to penetration or indentation (stiffness and temper)
- Ductility
  - Ability to undergo stretch and repeated bending cycles before deformation
- Burst
  - The ultimate strength test of a seamed pipe to rupture internally
- Collapse pressure
  - Deformation (crumble) of tubing from outside compression force

(Do you know the difference between high-carbon steel and low-carbon steel?)

***

Major Components of a Coiled Tubing Unit

1. Skid or flatbed mobile trailer (base frame for coiled tubing components)
2. Control console (where all the control gauges and levers are located)
3. Prime mover or the power-generating source

4. Coiled tubing string (various sizes of continuous pipe)
5. Hydraulic-operated steel reel (a large heavy-duty spool or metal wheel)
6. Coiled tubing measure counter (depth meter)
7. Level-wind assembly
8. Guide arch, known as the gooseneck assembly
9. Pipe injector head (to push or to pull the tube in and out of a well)
10. Pressure control equipment (stripper head and blowout preventers)
11. Hydraulic hoses and lines
12. Manifold valves and connections

Important Supporting Equipment Used in Coiled Tubing Operations

1. Triplex fluid pumps (high-pressure and low-volume rate pumps)
2. Supporting structure (high-reach crane unit and framed supporting legs)
3. Liquid nitrogen gas unit truck (cryogenic liquid nitrogen unit)

All of the above segments of coiled tubing will be briefly described as we pass through the book.

The liquid fluid pump, crane unit, and nitrogen cryogenic units are generally mounted on separate mobile trucks or skid units when they arrive at a well location. In offshore operations, we have limited horizontal space (platform or barge). Rigging up configurations will be different. Make sure you will have sufficient space at the well site to spot and rig up all the tools and equipment necessary to carry on with the task (onshore and offshore wells).

If you plan to work on a well with a small location, you may ask a coiled tubing company to provide tandem coiled tubing units, or you may have to expand the workspace. Tandem coiled tubing may be available for smaller jobs and smaller well locations.

If you have a workover rig on a well and require a coiled tubing unit, you may do so by rigging up the coiled tubing equipment cross-country to conduct additional wellbore interventions (sanded production tubing internally and externally). For example, sanding-up a well with a submergible pump (ESP) with tubing and electric cable could be a major fishing job.

The best choice you may have to use the coiled tubing is to clean up the tubing string internally prior to fishing operations. (You must have a sufficient area to rig up the coiled tubing unit and all the supporting equipment.)

Cross-county rig-up is a fancy oil field slang term. It means you may not have enough room to operate the coiled tubing along with the workover rig, and you must rig up the coiled tubing a distance away from the wellbore. If you are unable to rig up cross-country, you must rig down the pulling unit and deploy the coiled tubing equipment to conduct wellbore interventions first and/or rebuild and expand the location.

Tandem units (bobtail coiled tubing units) are unique in design and may save you the cost in rig-up and rig-down operations, a perfect setup for shallow low-pressure wells on land and/or barge work. On tandem coiled tubing unit packages, the fluid pump and the nitrogen storage tank are mounted on a single flatbed truck altogether. A small crane is mounted on the front of the coiled tubing unit itself, which will save you the time, money, and space that you may need. Tandem coiled tubing units are cost-saving packages that require less workspace to rig up and operate on shallow wells safely.

Coiled Tubing Operation
Versus
Wireline and Conventional Workover Rigs

Drilling rigs, workover well service rigs, wireline units, and coiled tubing each have something in common and are designed to carry out specific tasks (the choice is yours). The workover rigs, wireline operations, and coiled tubing are each designed to perform specific unique wellbore interventions. They may complement one another in certain tasks.

The wireline and remedial workover rigs (pulling units) are not part of this book, but the existence and application values of wirelines deserve mentioning briefly for your information. The tripping operation of a wireline or electric cable going into a well or pulled out of a well is safer and faster than the coiled pipe operation.

- The wireline must trip in and out of a well with the wellbore fluid in a static position under the protection of a pressure lubricator (no fluid flow during the trips).

- Workover rigs are generally slower operating units.

- Workover rigs will require blowout preventers and kill fluid to circulate through the tubing joints and to keep the well under control prior to tripping in or out of a well.
- The tubing string is wide open during tripping (you may get a kick at any time and be subject to cause environmental pollutions and accidents).
- Coiled tubing is capable of tripping in and out of a well under wellbore flowing pressure while circulating fluid (the well is always full of kill fluid).
- Coiled tubing is a constant well pressure control operation and is capable of conducting well interventions safely, effectively, and faster without killing the well.

I will fully present the application of steel wireline units and pulling units in detail in separate chapters later. The presentation of the workover rig and the wireline application are lengthy subjects to discuss in this paper.

The wireline can be a slick line or an electric wireline. A slick line is a single unbraided, shiny, smooth, and slick steel wire that is spooled or wrapped around a steel drum to store or transport to an oil and gas well for certain intervention operations similar to coiled tubing.

A slick line is a single long, continuous, un-welded wire available in different sizes: 0.108″, 0.125″, and 0.160″ in diameter or larger. Slick lines are manufactured in lengths of 18,000′, 20,000′, 25,000′, and 30,000′ to cover the deepest wells (check for temperature in deep wellbores). Tools are connected at the bottom of the slick line for specific downhole interventions. A slick line is not a conductor of electricity (it is not designed for certain wellbore interventions such as logging and perforating or correlations).

The electric wireline is designed to conduct logging and perforating in a well. An electric wireline is made of strands of wire cables (consisting of several wires bunched together with a core wire). An electric wireline is wrapped around a steel drum to be stored or transported to an oil or gas well to conduct various intervention operations similar to coiled tubing. Tools and equipment will be made up at the bottom of the electric wireline to run in and out of the well.

Slick lines and electric-cable wirelines are referred to as wireline operations. You may need to be more specific regarding which one you are going to use. Both the slick line and the electric wireline are spooled and mounted on a truck and/or skid, which is referred to as a wireline unit, similar to the coiled tubing spooler.

- The slick line or electric-cable wireline is fragile and depends on heavyweight bars and the force of gravity going down into a well under the protection of a pressure lubricator (static fluid in the borehole)
- You cannot run a wireline in a well without tubing/casing lubricator under the wellbore flowing pressure.
- You cannot use a wireline to circulate, drill, or mill objects within a well.
- A wireline cannot rotate going in or coming out of a well.

The electric wireline/slick line may be used for the following:

- Logging (temp logs, bond logs, collar logs)
- Perforating (through tubing or through casing)
- Sounding the wells (to tag and check well depth and catch samples)
- Bailing sand and getting solid samples
- Spotting cement using cement bailers
- Running and setting bridge plugs
- Cutting paraffin
- Cutting tubing strings (jet and/or chemical cuts)
- Backing off tubing/tubular strings
- Running and setting mechanical isolation packers under pressure
- Running and setting squeeze tools
- Production logging (finding out water channeling)
- Temperature survey and casing/tubing inspections
- Inclination survey

The application of coiled tubing, however, offers more of the same applications and advantages. Hydraulic force will push or pull the coiled tube in and out of the well, and it does not depend on the force of gravity to run the tools in or out of a well.

The most unique and valuable feature of using coiled tubing is the ability to wash down and circulate fluid going in the well and/or circulate oil and/or gas coming out of the well under a significant wellbore pressure or static well pressure (very unique engineering advantage). Most often, by using coiled tubing in a well control technique, you will find that the coiled tubing will provide a significant alternative advantage compared to other methods of well intervention solutions (a piece of advice).

Rigging up, rigging down, and fast tripping in and out of a well safely gives the coiled tubing a great advantage over the wireline operation and the conventional workover rigs that may take several days to complete the same job.

How Coiled Tubing Works

Continuous coiled tubing will be inserted into the production tubing and casing string under pressure while washing and circulating the well to achieve the intervention solution faster than conventional rigs.

Application of Coiled Tubing Unit in Oil and Gas Operations

Coiled tubing is a quick and cost-effective intervention solution without shutting or killing the well. Coiled tubing can be used in hostile oil and gas wellbores (high pressure, low pressure, hot, cool, deep, and shallow oil and gas wells).

1. Coiled tubing backwash cleaning outline
   - Oil wells offshore and on land
   - Gas wells offshore and on land
   - Flow line cleaning offshore operations
   - Injection wells
   - Saltwater disposal wells
   - Through tubing gravel packing operations (without pulling production string)
   - Jet washing and dewatering

2. Cementing service
   - Squeeze cement offshore and on land
   - Plug and abandon wellbores offshore and on land
   - Set balance cement plugs offshore and on land
   - Through tubing/casing treatments
   - Spot chemicals in a well offshore and onshore

3. Drilling practices (cost-effective measures to drill horizontal wells using coiled tubing than conventional drilling rigs to save tripping times)
4. Fishing objects while circulating and keeping well pressure under control
5. Milling operation (milling cement plugs, cast-iron bridge plugs) using motors

6. Well control measuring at all times, going in and pulling out of the well
7. Logging and perforating (while the well is under pressure)
8. Gravel packing (through tubing work)
9. Fracturing treatments
10. Subsea and deep wellbore cleaning (major cost-saving practice)
11. Effective and efficient pipeline cleaning operations offshore and on land

***

Major Components of the Coiled Tubing Unit

1) Skid and Base Frame

This is made of a heavy-duty welded steel I-beam as the base frame for the major components of coiled tubing. Heavy-duty axels and tires are mounted on the base beam.

The crew cab, control console, prime mover, hydraulic reel, goose neck, injector, and stripper assembly and all the well control stacks are built or mounted on the flat mobile truck trailer.

2) Coiled Tubing

As described in previous pages, the coiled tubing is a continuous length of flexible low-carbon steel pipe wrapped or spooled on a large steel reel for storage and/or transportation to oil/gas wells.

Coiled tubing is manufactured in a wide range of sizes. The most popular coiled tubing string sizes in the oil and gas business may range from 1″ to 1 ¼″, 1 ½″, 1 ¾″, 2″, 2 ⅜″, 2 ⅞″, and 3 ½″. Coiled tubing is manufactured in a wide range of grade sizes: 70 grade, 80 grade, 90 grade, 100 grade, 110 grade, and 120 grade

Coiled tubing is available in various sizes of long, continuous, flexible tubing without any tubing collars (no tool joints to leak or come off). Coiled tubing is manufactured as a seam type (ERW) and appears smooth and slick from the outside with no visible welding marks (you can see the welding seam from the inside of the pipe only).

Coiled tubing is mostly used for interventions in oil and gas wells. Some benefits of coiled tubing are noticeable and superior to wireline or slick line working operations. Coiled tubing has lower operation limitations compared to tubing joints or drill strings on the workover rigs or drilling rigs.

The coiled tubing limitations are related to the following:

- Tubing wall thickness
- Tubing outside diameter
- Tensile strength
- Chemical and mechanical property
- Corrosion resistance
- Erosion during well interventions
- Rotation abilities
- Well control calculation
- Patch and welding using butt weld
- Reduction of wall thickness and tensile strength
- Under constant fatigue forces

Coiled tubing cannot turn or rotate from the ground surface. The application of mud motors with different combinations of tools can be added below the coiled tubing to enable and accommodate washing, circulating, drilling, and performing other intervention activities.

***

3) Coiled Tubing Reel (Large Wheel)

The service reel is one of the most important pieces on the coiled tubing unit. The hydraulic power–operated service reel is used to spool the continuous flexible coiled tubing for storage and/or transport under potential tension force. (Do not let it unwrap!) The tubing reel provides a constant tension (back tension) on the pipe while storing or in motion, going in or coming out of the hole.

During a well operation, the service reel is designed to unspool the pipe while going in the hole and spool up the tubing while coming out of a well. The service reel is designed to hold, store, and/or transport several thousands of feet of uninterrupted, continuous, flexible slick steel tubing (with no tool joints). The service reel is designed to control the pipe between the reel and the injector system only.

The service reel alone will not have any power to push or to pull the pipe in and/or to pull out the tubing string by itself (its function is primarily limited to holding the pipe under tension and feeding the pipe to the injector head).

The service reel is equipped with a rotating swivel joint to pump fluid through the coiled tubing while the reel is in motion or standing still. The swivel joint is built on the reel shaft (unique design). In the offshore operation, the rig-up configuration may be different from the onshore installations.

The fluid-circulating manifold is part of the hydraulic reel assembly. Circulating connections are attached to the side manifold on the service reel. One line is used for water or chemical fluid injection, and the second line is used for nitrogen gas, where both fluids comingle or mix and enter the coiled tubing string (while the reel is rotating or standing still).

The high-pressure mechanical swivel on the drum axle of the reel drum enables the circulation of fluids at any coiled tubing position (while rotating or at rest). The high-pressure swivel gives the operator a choice to select different circulation installations if necessary.

Service reels have limited loading capacities based on tubing size, tubing length, and weight limitations (coiled tubing should be without any liquid to reduce extra weight while transporting the unit; some coiled tubing strings may hold as much as twenty-five barrels of water based on the size and the length).

4) Level Wind

The level wind is one of the components of the service reel. It is designed to align the coiled tubing as the pipe spools over the drum or off the drum (acting as steering and controlled from inside of the console).

The level wind consists of a level-wind guide that is operated by a lead-screw system. The level wind travels from side to side, aligning the tube to avoid coiled tubing crossing and crushing damage when spooling on or spooling off the reel (never align the coiled tubing by hand bars). The level wind is controlled from inside of the control console by the coiled tubing operator and can be raised at different heights.

The level wind holds the coiled tubing counter or depth meter also. The depth meter is to measure the length of the tubing going into the well or coming out of the well (very important indicator to measure and show pipe depth at any point while tripping in and/or coming out of a well).

The depth indicator must be checked regularly to prevent depth mismeasurements. When spooling the tubing out of the well, check the depth meter on the coiled tubing for mismeasurements.

Pinch points and slips mark damages on the coiled tubing. Depth meters are often found to be 10–100′ off depth on some coiled tubing units (avoid depth meter errors). A 100′ error to target depth is significant in some wells with short rat-hole intervals.

The coiled tubing lubrication is also controlled by the level-wind system (hand lubrication of the coiled tubing during the operation is not done correctly on most coiled tubing units). You need to lubricate your pipe as best as possible to avoid rust and corrosion, especially on the coiled pipe for wells of 12,000′ or deeper (the bottom layers of coiled tubing are subject to rust/corrosion because of the lack of proper lubrication).

5) Coiled Tubing Unit Braking System

The coiled tubing reel is equipped with a braking system to prevent the reel from uncontrolled rotation (because of hydraulic driving system problems). The brake system is used to prevent reel take-off without control. When the coiled tubing is completely wound up on the reel, the end of the coiled tubing must be kept in tension and secured with the brake system.

The stored potential tension and the elastic energy of the coiled tubing on the reel may tend to release and cause the unwinding of tubing, creating nests and damages (unpleasant scene). Protect yourself from the potential energy stored in the coiled pipe wrapping. Sudden unwrapped coiled tubing may cause accidents and is difficult to stop. (The pipe may shift in every direction to release the stored potential energy.)

6) Tubing Arch/Gooseneck

The gooseneck arch is used to guide the coiled tubing into the well. The coiled tubing will travel from the pipe reel under tension onto and over the gooseneck (similar to a guided channel). The arch consists of several control rollers mounted on the arch to assist bending, arching, shaping, and straightening the coiled tubing as uniformly as possible before the coiled tubing is pushed in, slid and snubbed into the injector blocks.

The gooseneck arch is designed to fit the coiled tubing outside diameter. The guide arch is an expandable solid piece of metal supported by the injector assembly and designed to collapse down in size when rigging down for transportation. The outside welded rod blocks over the gooseneck are designed to prevent the pipe from jumping out of the gooseneck path during buckling (when you're tagging obstruction or feeding the coiled tubing into the injector too fast, you will see the buckling effect on the pipe).

When the coiled tubing is spooled off the drum, the tube will be passed over the gooseneck assembly and through a series of steel rollers before getting into the injector without crushing the pipe. The main function of the gooseneck arch is to guide and slide the coiled tube over a series of rollers and force-feed it into the injector assembly.

The gooseneck arch radius is designed based on the size of the coiled tubing and API recommendations. Gooseneck arches may range from 70″ to 120″ in radius. Larger coiled tubing strings will be mounted on larger radii of gooseneck arches. Smaller tubing will be used over smaller arches (more flexible).

The application of correct-sized tubing and gooseneck radii is very important to minimize the effects of fatigue and oval shapes (egg shapes) on the tubing string as the tubing cycles (push or pull) over the arch. Smaller coiled tubing strings are fragile and weak to stand against too many applied fatigue forces.

Most coiled tubing failures are plastically deformed because of normal use over the gooseneck area. The coiled tubing fatigue failure is due to changes of loads when the coiled tubing travels up and down over the gooseneck assembly repeatedly. The misuse and misapplication of coiled tubing strings by the operator will cause deformation and holes in the pipe. High-temperature wells and acid-treated wells will cause corrosion and the loss of identity on the pipe.

Repeated high stress and aging will make the pipe thinner and susceptible to oval shapes, holes, or breakdowns. Repeated up and down motions of electric wirelines over the sheaves will also cause parting the wireline. Repeated motions of the coiled tubing over the gooseneck will be similar to the wireline and may cause serious coiled tubing deformation.

***

During my forty-two years of practical experience using coiled tubing, only four coiled tubing failures occurred on my job:

a. First Incident (coiled tubing in a low-pressure gas lift well)

While drilling and washing with the mud motor at a depth of 7,400′ at the plug back depth (PBDT), the 1 ¼″ pipe parted between the gooseneck arch and injector. The operator stopped the reel movement immediately. He bled off, spliced, and spooled out the coiled tubing safely.

b. Second Incident

During a saltwater disposal cleanup operation, using 1″ coiled tubing, we found a hole in the coiled pipe over the drum while going into the well. The operator stopped the operation and bled off the reel pressure. POOH (pull out of the hole) with coiled tubing and rig down (we exchanged the coiled tubing unit the next day to finish the job).

c. Third Incident

While washing and circulating mud, sand, and solids in a 2 ⅞″ production string at the depth of 6,430′, the 1 ½″ pipe was stuck. We were unable to move the pipe up and/or down (the cause was found to be fracture balls and a sand bridge). We sawed out the pipe above the ground and jet-cut the coiled tubing at 6,233′. We spooled out the cut coiled tubing and fished out the rested of the coiled pipe.

d. Fourth Incident (coiled tubing operation in low-pressure oil well [well sanded up])

In 2019, we had just started going in the hole and filling the coiled tubing with nitrified water at the depth of 500′ when the 1 1/4″ coiled tubing developed a large hole above the gooseneck (nearly parted). The operator stopped the reel immediately and shut down the liquid pump and nitrogen.

We cleared the personnel and waited on the nitrogen and saltwater to bleed off through the hole. We clamped the pipe on the drum side to make sure it did not run away. We cleared off the crew and worked the pipe a few feet to break the pipe (the pipe parted safely). We pulled part of the pipe out of the well using the injector, laid down 500′ of pipe safely, recut the pipe, and started in the hole the next day to finish the job.

In oil field operations, early detection and quick reaction to equipment failure and/or well pressure is a great disaster prevention. Most coiled tubing failures take place above ground level (the space between the Christmas tree and the coiled tubing reel drum).

Most Problematic Points of Coiled Tubing above the Wellhead

a. Parted coiled tubing between the gooseneck arch and the service reel
b. Parted coiled tubing between the stripper head and the injector
c. Buckled coiled tubing between the stripper head and the injector because of the snubbing force
d. Leak in coiled tubing between the gooseneck assembly and the drum reel
e. Leak in coiled tubing between the gooseneck and stripper assembly
f. Failure of coiled tubing above the ground or below the wellhead (must be carefully evaluated to avoid major well controls)
g. Parted and/or burst coiled tubing above the ground under the oil and gas wellbore pressure (will be difficult to control and must be evaluated and stopped quickly to prevent a major catastrophe, to protect everyone at the job site, and to prevent surface ground pollution)

In this case, the well cannot be killed in a timely manner and is deemed to become dangerous with time. The wellbore pressure attempts to go wild and dangerous (it will leave you with unpleasant and tough choices):

- Call for help at once.
- Quickly shut down any fluid circulation (fluid pump).
- Stop pipe movements immediately (stop drum movements).
- Close slip rams, followed by shutting the pipe rams and shear rams to cut the pipe.
- Consult with company man and your supervisor before jumping ahead and cutting the pipe with shear rams.
- Pick up on the sheared-off coiled pipe and close the blind rams.
- Prepare the kill sheet to kill the well and fish the coiled tubing out of the wellbore.

***

7) *Coiled Tubing Injector Assembly (Workhorse of All Components)*

The major coiled tubing force is carried out by the injector head and not the coiled tubing reel. As the name implies, the coiled tubing injector is designed to force in and/or to retrieve the coiled tubing out of the well by using hydraulic chain-linked gripper blocks (friction drive principle).

Hydraulic force will make two sets of double-row counter-rotating chains make the gripping blocks go up or down while holding or releasing the coiled tubing in constant motion and/or at a standing-still position. The basic operation of coiled tubing going in or coming out of a well is based on the friction drive principle.

The injector speed and directions are controlled by hydraulic force from inside the control cab. The weight indicator on the coiled tubing is similar to the workover rig. The hydraulic load cell bladder will transmit the heavy pipe and/or light pipe to the operator. The weight indicator is very important in coiled tubing operations (may be hydraulic and/or electrical indicators on coiled tubing units).

The hydraulic-driven tubing injector is a major part of the coiled tubing unit operation. The hydraulic-powered injector will provide sufficient force to grab, pull, push, or bend the pipe over the gooseneck and run the coiled tubing string in and out of the wellbore without damaging the pipe. The injector is the power behind pushing in and pulling out the coiled tubing (snubbing and stripping).

The major part of the coiled tubing unit work is carried out by the injector head. The injector head is the actual workhorse that applies necessary hydraulic force to the tubing string while going in the well or pulling the entire coiled tubing string out of the well at desired speed.

The applied force directed to the coiled tubing is actually derived from two counter chain-links powered by hydraulic motors with sets of circular steel gripper blocks that are attached to the chain-link to continuously grip and/or release the pipe while in motion or standing still (with no visible pinching or slip marks on pipe).

The injector head may have an automatic braking system that activates in case of a coiled tubing "runaway" situation and/or drive motor problems (constant inspection of steel gripper blocks is necessary to avoid jamming).

Major Functions of Injector Head Assembly

The coiled tubing injector is the main force behind the pushing and/or pulling the coiled pipe going in or coming out of the well.

- Snub the pipe in the hole against wellbore pressure (push the pipe in).
- Apply force during milling and drilling operation.
- Snub the pipe to overcome sliding and friction.
- Hold and control the speed of going in and coming out hole
- Hold and support the entire tubing under load during operation
- Pick up and strip the tubing out of the wellbore.

***

8) *Control Cab (Console)*

The control cab is a small built console where the coiled tubing operator manages, monitors, operates, and controls all the hydraulic and pneumatic coiled tubing components (levers and gauges).

The purpose of the control console is the following:

- Monitor operation of the coiled tubing movements while going in the hole or coming out of the hole
- Avoid damage to hydraulic gauges
- Keep monitoring and maintaining of the gauges to insure they are dry and free from mist, rain, and dust
- Reduce outside equipment noise while working and communicating
- Have a controlled central operation point
- Act as the command communication center

There are several important gauges and levers that are located on the panel of the cabinet that must be monitored during operation:

a. All blowout preventers (close, open, BOP system)
b. Hydraulic system (power pack control)
c. Levers to service reel, injector system, and level-wind assembly
d. Pressure on coiled tubing string (injection pressure)
e. Displacement volume (filling and displacement volume)
f. Coiled tubing running and pulling speed
g. Depth counter (operating depth)
h. Tubing weight (tension or compression)
i. Outside noise (keeps down)
j. Wellhead pressure
k. Data gathering and reporting

***

9) Power Unit (Prime Mover)

The power unit is critical in the operation of important components of the coiled tubing. The power unit is often provided by diesel engines.

There are several sources of power supply on coiled tubing operations (power pack):

- The power to (control console) operate the service reel
- The power to the injector
- The power supply to blowout preventers, generally the hydraulic actuated system
- The power to operate the panel valves provided by a pneumatic and/or hydraulic system

Some power generators or prime movers are used on some units. The coiled tubing unit is run on either power pack (diesel engine, hydraulic, and/or pneumatic).

10) Fluid Pumps

The fluid pump used in coiled tubing unit operations is a high-pressure and low-volume positive displacement pump. The pumping rate through the coiled tubing depends on the size of the coiled tubing string. Smaller tubing strings will have a lower rate and higher pressure because of the small diameter and long continuous pipe (fluid friction). The fluid pumps in coiled tubing operations are usually small high-pressure and low-rate triplex pumps (three-plunger, four-plungers, or five-plunger pumps).

The fluid pump must be in good operating condition before application. For example, a normal pumping rate through a 1 ¼″ pipe may range from a ¾ barrel to a maximum of one barrel per minute. The higher the pumping rate, the higher the pumping pressure (do not burst the pipe). Coiled tubing has a limited rate and pressure (stay within the safe pressure margin to protect the pipe from deformation).

***

11) High-Reach Supporting Crane Unit

The supporting structure of crane units are referred to as high-reach cranes or cherry pickers with adequate lifting power and reaching capabilities. Cranes are used to lift and install well control equipment, support, and hold tension on the coiled tubing assembly to avoid the heavy weight of the injector and the BOP slack-off the wellhead or the tubing joint above the wellhead (Larkin-type isolation wellhead).

The purpose of a crane mast is to hold tension and support the gooseneck, stripper head, injector, and blowout preventer equipment from swinging side to side and also reduce weight to a minimum or no weight on the wellhead equipment.

Stable support is necessary to hold tension on the coiled tubing equipment above the wellhead to avoid potential damage to the wellhead and keep the injector system from moving or falling. Other types of structural support are available for offshore or land operations. Below the injector assembly are usually four aluminum pipes acting as drop legs that can be utilized for additional support.

***

12) Application of Nitrogen (N) in Wellbore Cleaning

Application of nitrogen (N2) in coiled tubing intervention operation has many impressive benefits. The foam application nitrified acidizing, nitrified water in the wellbore cleaning and stimulation are only a few benefits to an oil and gas field.

Liquid nitrogen is stored or transported to oil and gas wells in special cryogenic stainless steels tanks, for safety purposes. The cryogenic stainless steel tanks consists of the inner and outer shell tanks which is capable to withstand extreme cold temperature of -320 degrees and compressive gas pressure.

The cryogenic liquid gas tanks are designed with pressure gauge and control valves to bleed gas expansion pressure in order to avoid accidents. Operating the liquid N2 requires a trained person with complete personal protective equipment (PPE).

- nitrogen can burn skin tissue
- nitrogen gas cloud will reduce oxygen
- loud noise when bleeding off nitrogen

Our atmosphere consists of oxygen, nitrogen, hydrogen, helium, water vapor, argon, and carbon dioxide elements. Normally, the earth's atmosphere is made up of approximately 78% of nitrogen.

Nitrogen is nonflammable, nontoxic, colorless, odorless, noncorrosive, and safe to work with. Nitrogen alone is a very safe gas to work with. It will not burn or explode. Actually, you can use nitrogen to put out a fire if available.

The application of nitrogen in the oil field began in 1956 with the use of a nitrogen gas cushion in a drill stem testing operation. Nitrogen application has been expanded to a larger market in oil and gas, including fracturing foam and continuous applications in the wellbore cleaning (comingling of nitrogen with water).

Nitrogen does not react with other elements and is slightly soluble in water, oil, and other liquids. Liquid nitrogen gas is transported to well locations in cryogenic storage tanks safely for the purpose of wellbore cleaning in offshore and land operations.

Caution: Never use free air in the foam unit for wellbore cleaning applications. Never use dry air to jet and clean up any oil and gas wellbores.

- The application of dry air may be dangerous in wellbore cleaning.
- A chemical reaction of dry air with hydrocarbon products may cause explosions and/or underground blowouts.
- Introducing free air and oxygen will increase corrosion and bacterial growth in the wellbore.

The application of nitrogen when mixed or comingled with water (nitrified liquid) is very safe, effective, and efficient fluid to clean up and circulate sand or solids out of a wellbore with great success ratios.

Liquid nitrogen is transported into a special cryogenic liquid nitrogen tank under pressure. A cryogenic liquid nitrogen tank is built on a flatbed mobile transport truck. At the liquid stage, nitrogen is compressed to approximately −320 degrees Fahrenheit. (LN2). While nitrogen is a nonflammable gas, liquid nitrogen is extremely cold and can burn you. The nitrogen unit is designed to pump the liquid nitrogen from the cryogenic tank to a special converter that heats up and changes the liquid phase to a dry nitrogen gas.

The nitrogen gas product has many useful applications worldwide, including but not limited to displacing wellbore liquid in oil or gas wells to help the well flow naturally. Normally, nitrogen comingles with water to assist lightening the heavy wellbore fluid and to reduce hydrostatic fluids circulating abrasive solids out of the wellbore (it will avoid the loss of circulation).

Fresh water is 8.33 pounds per gallon, and some reservoirs produce 9.2 pounds per gallon of saltwater (too heavy to circulate without comingling with nitrogen, 0.4784 measured fluid gradient in some wells). The application of nitrogen in water-sensitive wells is a good practice to prevent water from entering the reservoir

***

13) Steel Tanks/Reservoir Pits

Steel pits are required during coiled tubing operations to circulate fluids into the pit. One or two steel tanks are necessary to keep circulating fluids from spilling over and contaminating the ground (make sure the steel pit is clean and does not have a hole at the bottom of the steel tank). I do not recommend using a dirt pit for coiled tubing wellbore cleaning. It is a bad practice and should be avoided.

14) Vacuum Trucks

Normally, two vacuum trucks are necessary to assist the coiled tubing operation: one truck to supply clean fluid for the coiled tubing fluid pump truck and the other to haul dirty water in case of an emergency. Do not run out of water to the coiled tubing while washing and circulating; it may get the coiled tubing stuck. Do not allow the backwash tank to spill over and cause pollution.

15) Pressure Control Equipment and Blowout Prevention Stack (BOP Stack)

The impressive part of the coiled tubing is the pressure control equipment (gets the job done safely).

Did you know that most often, a well will communicate to you before a kick (sending signals that an influx is moving up the hole)? The word *kick* is an oil field word that means the sudden entry of a high-pressure formation influx coming into the borehole/wellbore (*influx* means high-pressure oil, gas, and water).

The pressure control equipment is designed to hold and control high and low wellbore pressure during coiled tubing operations if needed. It is very important that all the pressure control equipment and blowout preventers meet or exceed the API requirements when working on any oil or gas well (follow the API recommendations).

All the well control lines, valves, chokes, and manifold connections must be checked and tested to maximum anticipated working pressure (MAWP) prior to the coiled tubing operation. Never assume that a well has no pressure (every single oil and gas well must be considered a live well).

All the equipment must be maintained in good working condition before working on any producing oil and/or gas well. The pressure control equipment must be spaced and placed below the coiled tubing injector and above the wellhead. (It must be easy to reach them in case of emergency repairs.)

The coiled tubing pressure control equipment may consist of primary and secondary control equipment.

The coiled tubing stripper head assembly (primary control) is the most important component of the coiled tubing unit's safety operation features.

The stripper pressure control is often called the pack-off and/or stuffing box. The stripper head has a similar function as an oiler on swab lubricator or the wireline pack-off assembly. The stripper head is the first active well control component during a coiled tubing operation (it is the first and most important line of defense).

The packing element must be improved drastically to avoid failure during well control measures. Did you know that the stripper head is the first isolation preventer that keeps you safe during actual coiled tubing work? (The stripper head must be highly dependable.)

The stripper head and pack-off element are located immediately below the coiled tubing injector head and above the coiled tubing BOP stack.

The purpose of the stripper head pack-off is to isolate and contain wellbore pressure below the tools and prevent high/low-pressure fluid from leaking or leaching out around the tools from the annulus to the surface while going in, coming out, and/or being in a stationary position during an intervention operation.

The annulus is the space between the outside diameter of the coiled tubing and the inside diameter of your production tubing (or casing string, whichever it may be).

The pack-off elements must be of a correct size and dependable and create a positive seal around the coiled tubing during the operation while circulating, going into or coming out of a well and/or in a stationary position. I do not like or trust the pack-off elements in the stripper head (need to improve this device).

Pressure control on the coiled tubing must be the first priority before any oil and gas wellbore intervention planning to protect people, the environment, and properties. All the well service operating rigs or service units must provide dependable and adequate well control equipment before entering any oil and gas well.

- A coiled tubing unit operation must be equipped with high-quality and dependable stripper head assembly and a complete BOP stack consisting of but not limited to the following types of ram preventers: blind rams (top), shear rams, slip rams, and pipe rams in case of emergency (at the bottom).

- Stripper head is a very important tool in coiled tubing operations. In some coiled tubing operations an annular preventer may be used. Annular preventer's may be a preferred selection when working with long coiled tubing and multiple BHA of various outside diameters.

- Electric wireline conducting: logging, perforating, setting bridge plugs, running and setting isolation tools, cement retainers, cement bailers, or any other downhole wellbore activities must be equipped with adequate well pressure control equipment such as casing or tubing lubricators with hydraulic pack-offs and/or grease injectors for anticipated high wellbore pressure. Lubricators must be volume-tested to maximum anticipated wellbore pressure (leak free).

- Slick line unit conducting: cutting paraffin, bailing sand, perforating, fishing objects, swaging tight spots, or even tagging well depth must be equipped with complete pressure lubricators and hydraulic pack-offs and BOPs in case of well flow emergency (you're never going to know after you break through a bridge).

    - Do not use slick line as a swab unit. A slick line is not designed to swab a well. (It is not a good or a safe practice.)
    - Do not use a slick line to spot chemical or acidizing
    - Do not use a slick line unit to perforate tubing and/or casing.
    - Do not run pressure bombs or well surveys after acidizing a well.

- Swab units must be equipped with isolation valves and a BOP below the swab lubricator. Always rig up on a crown valve/swab valve above the master valve (do not rig up on a master valve). The swab lubricator must be rigged up with a hydraulic oiler saver (double) to contain oil and gas pressure during the swabbing operation (read the book *Field of Swabbing* by this author).

***

The selection of coiled tubing pressure control stripper depends on the anticipated wellbore operating pressure (never underestimate wellbore pressure). Using a coiled tubing string in major well control problems is a concern issue.

There are several types of strippers available for coiled tubing work. Good-quality packing elements are critical to avoid pack-off failure during coiled tubing operations. Stripper packing is the first and most important isolation that must be taken care of (your first line of defense against high wellbore pressure is the consumable rubber elements).

On high-pressure coiled tubing interventions, the coiled tubing may have two strippers with additional BOP components. Regardless of types of coiled tubing and/or wireline well intervention operations, all the insert packing elements are consumable and must be inspected and exchanged regularly. (Never shortcut or take a chance when it comes to safety practices.)

The pack-off elements can be changed-out in coiled tubing side-door strippers more quickly and easily than the conventional top-entry strippers. (It would not take but a few minutes to exchange the packing element in the side-door stripper.) Check the coiled tubing regularly. Oval-shaped (egg-shaped) pipes will cause fluid leaks.

***

Coiled Tubing Blowout Preventer (BOP) Stack

The blowout preventers on the coiled tubing operations are normally secondary selections on coiled tubing well control (the blowout preventers are there for high pressure and emergency purposes only).

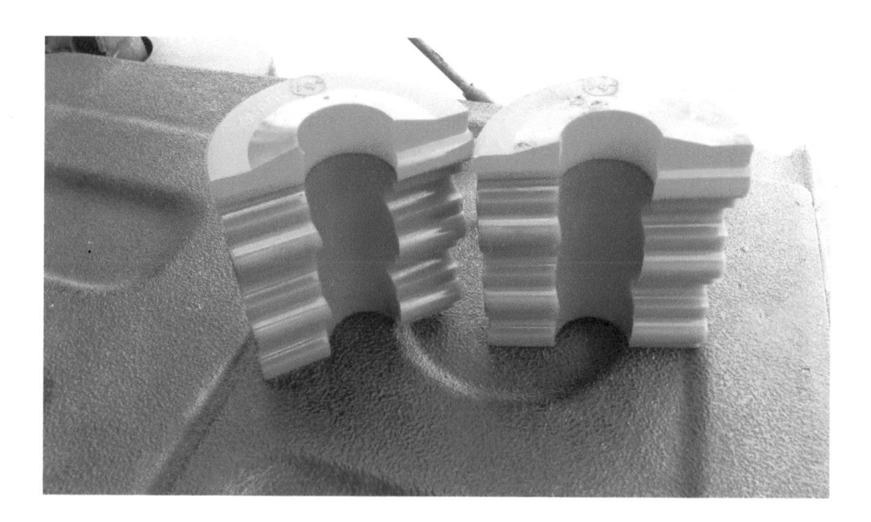

The blowout preventer stack consists of the following:

- The blind rams (on the top to seal off the borehole)
- The shear rams (to shear or cut coiled tubing)
- The slip rams (to hold the pipe from falling)
- The pipe rams (at the bottom to seal off around the pipe)

1. Blind rams are used to close the wellbore when the coiled tubing string is out of the hole. The purpose of a blind ram is to close shut the wellbore and prevent fluids from flowing out of the well while the tubing is pulled completely out of the hole. Blind rams are located on top of the BOP stack assembly.
2. The shear rams are used in case of an uncontrollable well condition or emergency (last chance to escape). The shear rams are designed to cut or shear off coiled tubing in order to close the blind rams or the wellbore in.
3. The slip rams are designed to grab hold of tubing and support the coiled tubing from dropping into the well or being pushed up the hole in case of unexpected emergency situations of parted pipes or shearing the pipe off.
4. The pipe rams, as the name implies, are used to close and seal off tightly around the coiled tubing from the outside and to isolate the annulus only. The annulus is the space between the inside diameter of the production tubing and the outside diameter of the coiled pipe (you cannot circulate or bullhead fluid down the annulus if the pipe rams are closed).

All the coiled tubing employees must be trained and familiar with the purpose of all the blowout preventers to avoid mistakes such as shutting the wrong set of rams and creating downhole well control issues and/or major fishing problems (has happened before).

The blowout preventers should be color-coded for clear identification. If you think the cost of training is high, wait until you see the cost of ignorance. Shutting blind rams and/or shear rams by mistake will cut or may cause damage to the coiled tubing by flatting the tubing—seriously.

***

Coiled Tubing Areas of Operation Worldwide

1. Offshore operations: Skid-mounted coiled tubing components are shipped to offshore operations across the world. Coiled tubing work on jack-up rigs or production platforms is restricted to limited weight and space.
2. Shallow bay water and rivers: Coiled tubing work on the wells located in shallow bay waters is normally loaded on flat barges and transported to the well site.
3. Land operations using truck-mounted coiled tubing units: The land coiled tubing units are generally quite mobile from field to field and easy to rig up and rig down.
4. Foreign countries: Some of the components of coiled tubing operations are shortcoming and often are difficult to find.

All of the following pertain to coiled tubing:

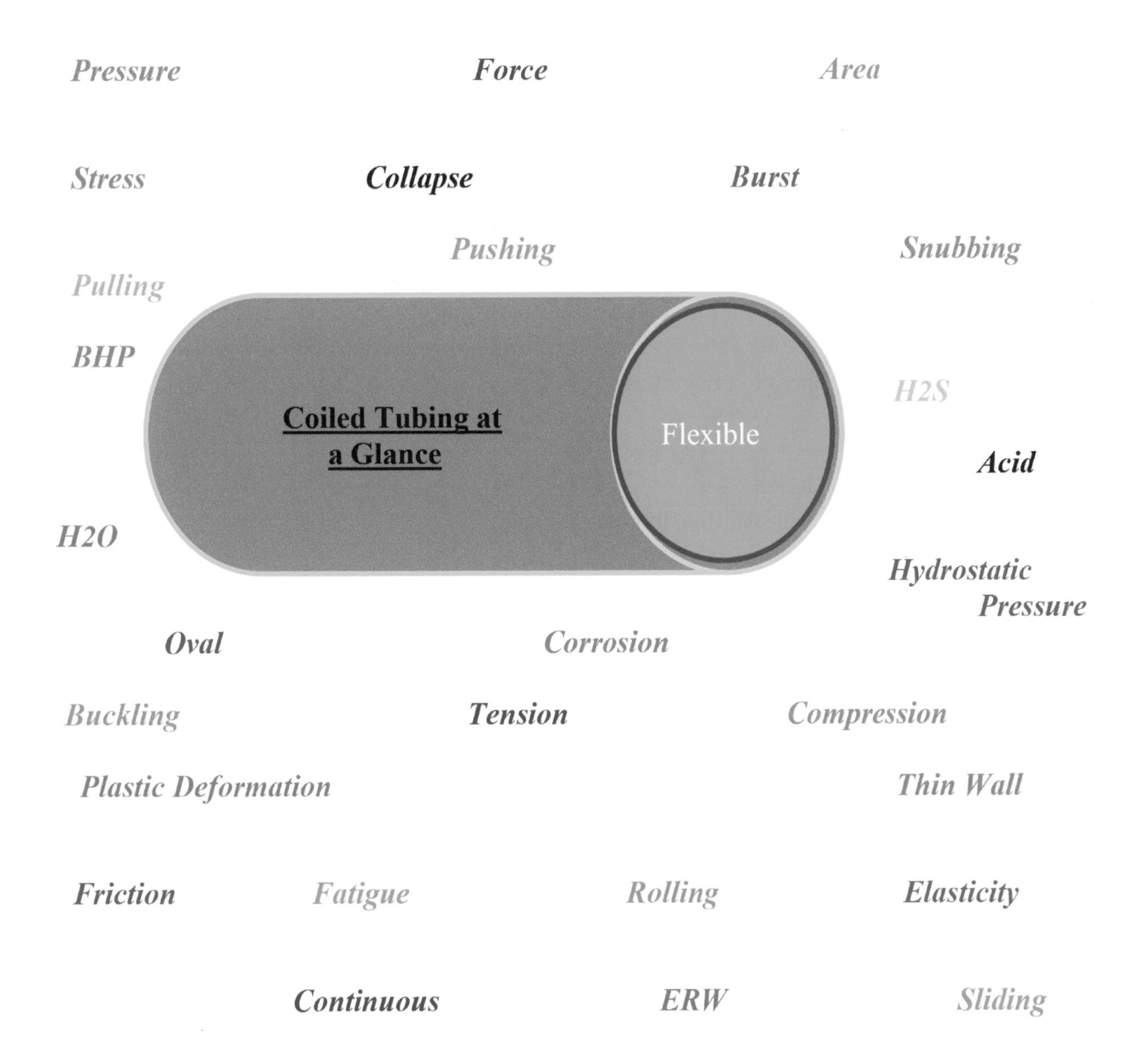

Operation: Coiled Tubing
What Do You Know about Well Pressure Control?

Using coiled tubing strings in major well control may be a concern.

- Keep yourself safe from well kick pressure.
- Keep the coiled tubing equipment safe and maintained.
- Keep the coiled pipe safe from stress and strains.

Basic well control pressure calculation formulas are necessary to work on any oil and gas wellbore intervention in the world (educate yourself and your people in well control). Well control formulas are the key to a safe wellbore operation regardless of well intervention method.

➢ Drilling operation
➢ Coiled tubing operation
➢ Completion of new or older wellbores
➢ Remedial workover operations

Coiled tubing application is primarily manufactured for downhole interventions and the well control solution. The safety and success of any wellbore operation depends on the quality and integrity of the coiled tubing components. Blowout accidents across the world, offshore and onshore, are mostly due to the lack of proper equipment and poor training.

- What is pulling?
- What is pushing?
- What is pressure?
- What is force?
- What is stress?

All of the above will apply to the coiled tubing in operation.

Mathematically, pressure means force over the area of a solid object.

- Pushing or pulling on a solid object may be called force or stress (putting weight on packer, jarring on tubing, or pulling on stuck pipe).
- Any act of pushing or pulling on fluid (gas/oil/water) is different and may be expressed as pressure (psi) (squeeze pressure, pump pressure, formation pressure, hydrostatic pressure, hydraulic pressure).

If you learn and understand downhole pressure, you will be ahead of the game. Most new engineers and managers must learn well control techniques.

In this writing, we will deal mostly with fluid and wellbore pressure.

$P = F/A$

P = pressure (psi)

F = force (pounds)

A = cross-section area (square inch)

$F = P \times A$

P = pressure (applied to fluid)

F = force (applied on solid)

A = cross-section area of an object

F/A (stress on an object)

When talking about pressure, we are usually talking about fluids. Well fluids can be differentiated as gas or liquid. Any fluid height will exert pressure (fluid filled in a stock tank, fluid filled in a vertical pipe, and/or fluid filled in a wellbore). The height of the fluid column at rest is referred to as the hydrostatic head or hydrostatic pressure (psi). The hydrostatic head refers to the weight of the static column of fluid at a specific point.

Hydrostatic pressure: $.052 \times$ height $\times$ fluid density

Hydrostatic pressure: pressure gradient $\times$ depth

Hint: 1 gallon of fresh water = one pound = 231 cubic inch

When we are talking about wellbore fluid in a well, we are dealing with the following:

- Hydrostatic pressure (psi)
- Pressure gradient (psi/foot)

The basis of the fluid pressure gradient starts from fresh water. The salinity of your drinking water is about 8.33 pounds per 1 gallon.

1 gallon of fresh water = 1 pound = 231 cubic inches

Hydrostatic pressure depends on the height and density only (it does not have anything to do with how large the pipe or vessel is). The formation fluid gradient is based on salinity or the density of the formation fluid from the ground level down the earth (30,000′ and beyond).

The deepest well I have worked on was 22,000′ (with 8,650 psig of pressure, 2,100 barrels of oil per day, with 200 MCF of gas and no water).

Pressure gradient (psi/ft.) = .052 × MW (pounds per gallon)

Hydrostatic pressure = pressure gradient × height (psi)

Hydrostatic pressure = .052 x height x density

What Is Formation Pressure?

Formation pressure is the pressure that is trapped in the formations of tiny pores (often called pore pressure). When a well is flowing, it means the reservoir pressure is higher than the column of fluid in the tubing string and/or casing from the reservoir depth to the surface. As the formation pressure declines naturally, the hydrostatic head will increase, and the well tends to die.

When the well stops flowing, it means that the formation pressure (pore pressure) becomes equal to the hydrostatic column of fluid in the well (oil, water, and gas) from the surface to the perforations. Low fluid level in the wellbore is the indication of a depleted reservoir or perhaps, plugged-off perforated holes in the casing.

Formation pressure and temperature are big factors to deal with. Some theorize that the structure or the origin of formation might come from displacement and deposit elements such as sand, shale, clay, gypsum, lime, and dolomite and may be the result of rain, rivers, and oceans (there is much more to it than that).

Drilling operations and excavations deep into the belly of the earth indicate that our planet we are walking on has been rumbled and rolled over several times and reformed to perfect shapes and construction as we see it. Every layer of the earth is built with significant trapped pressure from the surface of the earth down.

There are basically four elements in the reservoir to deal with. Reservoir formation is made of a dead element with special physical and chemical characteristics. Formation is the basic building block or trap for oil, gas, water, and others.

Oil, gas, and water will support the physical shape and condition of the formation as it stands. The formation will collapse without the support of underground water, oil, and gas pressures. Changing the condition of these elements will cause sand and solids to cave into the wellbore. Water, oil, and gas are either trapped in the same reservoir formation and/or separated by some sort of non-permeable formation block (shale break).

Based on each natural element's characteristics, oil, gas, and water are immiscible elements and tend to separate or position themselves based on given densities. Water stands at the bottom, with oil in the middle and gas on top (repel).

The contact point of each element is called oil and gas contact or water-to-oil contact. The contact points of oil, gas, and water are important to know for completion and perforating purposes (don't shoot in the water zone and do not deplete the gas cap). Fresh water is perfectly separated from saltwater; they do not cross each other's boundaries. Even in the open ocean, saltwater and fresh water do not mix (they reach but do not cross boundaries).

Mathematically, as we drill down the earth, it will indicate that pressure increases with depth. An increase in pressure can be calculated based on the salinity of the formation fluid, referred to as the gradient. The pressure gradient is measured based on the salinity of the trapped liquid in the formation's tiny pores.

We measured the density of fresh water to be 8.33 to 8.34 pounds per gallon. The 8.33 pounds is the density and/or weight of fresh water that you are drinking. Fresh water exerts pressure and is calculated below:

$0.052 \times 8.33 = 0.43316$ psi/foot of pressure (called gradient)

The reservoir pressure is calculated based on the formation's fluid pressure gradient. The pressure gradient of fresh water is 0.433 psi/foot, and the pressure gradient of saltwater taken from a shallow reservoir formation may range 0.465 psi/foot and may increase as we drill deeper (pressure and temperature will change with depth).

In pressure control, all the reservoir pressures ranging between 0.433 and/or 0.465 psi/foot are considered normal pressure gradients (from the ground surface to the deep wellbore intervals). Any formation fluid gradients higher than 0.465 psi/foot are considered abnormal pressure gradients (take caution when drilling through abnormal-pressure reservoirs). Some abnormal pressure gradients may reach as high as 0.985 psi/foot or even 1.01 psi/foot.

Calculating abnormal reservoir pressure at 6,000′ with fluid pressure gradient 0.985 psi/foot leads to the following:

$6{,}000' \times 0.985 = 5{,}910$ psi (considered as abnormal reservoir pressure)

Formation pressure may be caused by the overburden force (the compaction of the earth's crust). High overburden pressure will force and squeeze the sand grains tightly and exert pressure on the trapped fluid in formation pores, causing abnormal reservoir pressure. The overburden force is estimated to be 1 psi/foot.

All the reservoirs will have pressure. Some reservoirs will have low pressure, and some reservoirs will have significantly high pressure. You are seldom going to see abnormal reservoir pressure at a shallow depth. If you have 8,865 psi of trapped pressure at a shallow reservoir depth of 4,000′, we would be thrown up into the sky and into the oceans!

Have you drilled into an abnormal shallow saltwater reservoir (tough to control)?

Some companies ignore "not to" fracture shallow oil or gas reservoirs.

You may see the ground start to bubble, oil, gas, and saltwater 60′ away from the wellhead because of the fracturing pressure. Some people may be wondering why. (They may ask themselves, "What in the world has happened?")

I am against any shallow wellbore fracturing above 5,000′ (stop contaminating fresh water). Why are you fracturing a well that is on a vacuum? We need to teach and train people. If you have foreign and/ or domestic students in your class, please slow down before throwing all types of fancy words and formulas at them too quickly. Most of them do not have a clue what in the world you are talking about.

You are not reaching in your teaching.

The oil field business is flooded with many green graduate engineers who do not know what a wellbore looks like.

Hydrostatic pressure:

Hydro is the fluid. The word *static* means at rest (not flowing).

HP = .052 × TVD × MW

Example:

Perforation depth: 3,800′–3,850′ (TVD/total vertical depth = MD)

MW: 9.2 pounds per gallon

Shut-in DPP = 750 psi

Tubing: 2 ⅜″ EUE 8rd 4.70#

Casing: 7″, 23# at 6,500′

HP = (.052 × 3,800′ × 9.2) = 1,818 psi

BHP = 2,568 psi

Do you know what it takes to kill the above well? Calculate it for yourself!

Basic Important Formulas in Oil/Gas Well Control

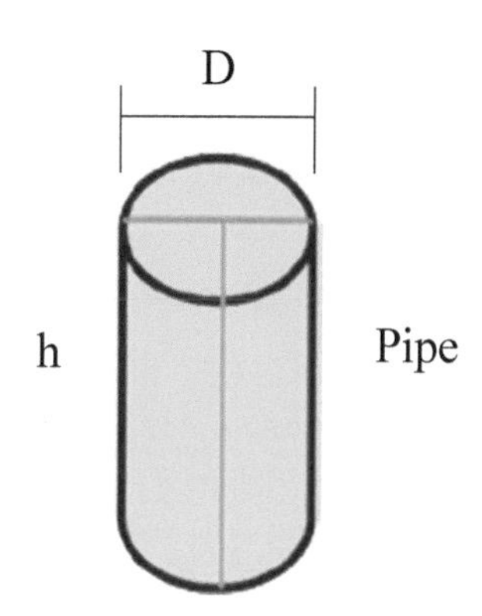

- hydrostatic pressure = 0.0519 × fluid height × fluid density
- area of tubing (circle) (pipe inside diameter)2 × 0.7854
- area of tubing (circle) (pipe radius)2 × 3.1416 [π]
- pressure = force divided by cross-section area
- hydrostatic head = the weight of column of fluid at rest
- R = radius of the circle
- D = diameter of the pipe
- Hint: 1 gallon of freshwater = 1 pound = 231 cubic inches
- 100 pounds of sand = 1 cubic foot
- 2,000 pounds = 1 ton
- 1 gallon = 3.785 liters
- All the oil- and gas-production reservoirs have pressure.
- Some reservoirs will have significantly high pressure, and some wellbores will have lower pressure.
- If the reservoir pressure is higher than the hydrostatic head of fluid in the tubing/casing, the well will flow.
- If the reservoir pressure is equal to or lower than the hydrostatic head of fluid in the tubing/casing, the well fluid will stay static at different heights below the surface.
- If you swab the well fluid, you will reduce the hydrostatic pressure, and the well may flow naturally.
- When you push the pipe into the fluid into the wellbore, you may raise the fluid level and the hydrostatic head, which may cause the fluid to spill or run over (this is called displacement).
- A high gas–oil ratio is dangerous and may cause unwanted fluid kicks. A kick is defined as the intrusion of a high-pressure reservoir influx into the wellbore. Influx is the amount of oil, gas, and water moving into the wellbore.
- A low fluid level in a well is indicative of low reservoir pressure, a depleted reservoir, and/or a sanded-up wellbore (perforations may be covered up with mud, sand, or solids).

Important Terms Used in the Oil Field

Capacity is "the ability to hold or ability to contain within." *Capacity* is "the volume of a certain tube to hold fluid [tubing, casing, and stock tank capacity]."

Displacement is basically "replacement measurement." (When lowering the tubing string into a well full of fluid, the excess water that spills out is the replacement or tubing displacement).

The volume of a fluid is equal to the mass of an object (tubing, tools, boats, and so on).

There is a significant difference between pipe displacement while going into a well and a natural well-flow pattern. (It is important to understand and remember that in order to avoid blowout accidents.) Always check by shutting down the well for fluid flow while tripping a pipe in or out of a well (early detection and a quick reaction).

What Volume Means

Volume is the measurement of physical space in a container in terms of width, height, expansion, or contractions. One cubic foot is equal to 7.481 gallons.

Differential Pressure

We use the term *differential pressure* a lot in oil field calculations. Differential pressure simply means the difference between the two pressure areas (for example, the differential pressure between the inside and the outside of a tubing string). The differential pressure below a plug and above a plug may be significant. Do not take differential pressure lightly (differential pressure is often misunderstood and/or underestimated, which may cause well blowout or serious accidents).

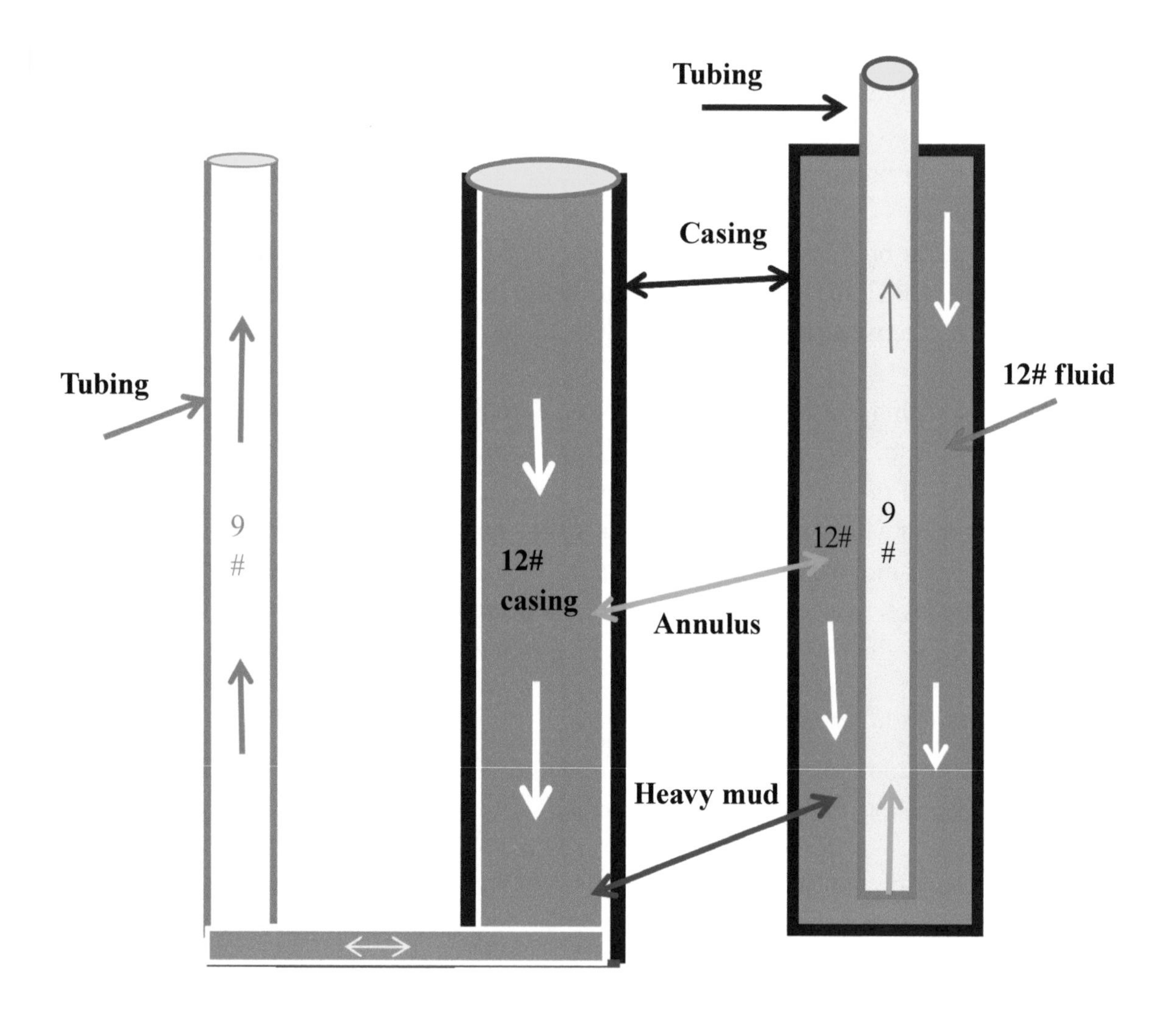

Basic Calculations You Need to Know in Oil and Gas Operations

Tubing Capacity (Plug in your numbers!)

Example: What is the capacity of 2 ⅜″ 4.70 lb./ft. tubing or pipe?

(2.375″ OD)

(1.995″ ID)

Barrel per linear foot = $0.0009714 \times D^2$

D = inside diameter of tube

Barrel per lineal foot = $0.0009714 \times (1.995)^2 = 0.003865$

Barrel per lineal foot of 2.375″ pipe = 0.003865 barrels = capacity of

2 3/8″ 4.70# tubing

Capacity per 1,000′ = $0.9714 \times (1.995)^2 = 3.87$ barrels

or 1,000′ × 0.003865 = 3.87 barrels

Linear feet per one barrel = $\dfrac{(1{,}029.4)}{D^2 = (1.995)^2} = 258.60$ feet

or 1 bbl. divided by 0.003865 = 258.60 feet

The capacity of casing or tubing (barrels per foot) = $0.0009714 \times (\text{inside diameter of pipe})^2$

Liter per meter = $\dfrac{D^2}{1273}$ Meter per liter = $\dfrac{1273}{D^2}$

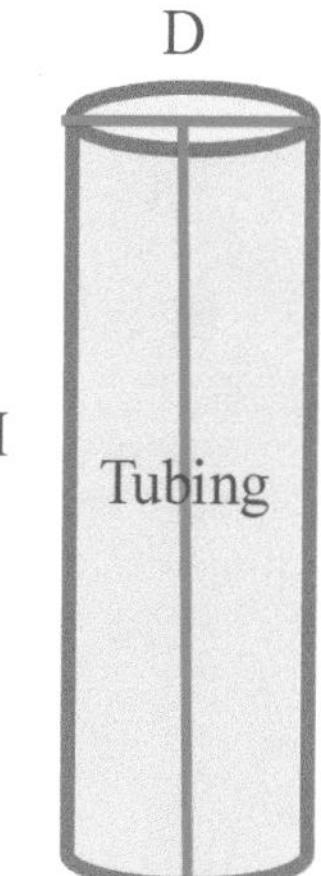

Learn the MKS system; it is more accurate than the CGS system.

Basic Compatible Kill Fluids in Coiled Tubing Operations

- Field-produced water (saltwater)
  Field-produced water is compatible to most low-pressure reservoir formations. Field-produced water may be produced in various weights, ranging from 8.6 pounds per gallon to 9.2 pounds per gallon.

- Sodium chloride solution (NaCl) 8.3 to 10 pounds per gallon

- Potassium chloride solution (KCl) 8.4 to 10 pounds per gallon
- Calcium chloride solution (CaCl2) 8.3 to 11.7 pounds per gallon (PPE is required)

  [Watch it! It will wrinkle your leather shoes, gloves, shirts, and pants.] (Read MSDS.)
- Drilling mud 10 to 17 pounds per gallon (PPE is required)
- You need MSDS sheets for CaCl2 or KCl fluid.

The Continuous ERW Coiled Tubing String

The coiled tubing wall is designed to be thin, light, and flexible with a larger inside diameter than the standard tubing joints in the oil field.

Coiled tubing is manufactured in the following grades:

- 70 grade
- 80 grade
- 90 grade
- 100 grade
- 110 grade
- 120 grade

Each grade of coiled tubing_has a special physical and chemical makeup, displacement, and capacity. The design and manufacture of the continuous flexible coiled tubing is different from standard oil field tubing joints. Make sure you learn about coiled tubing grade and dimensions during any oil and gas intervention. Use coiled tubing dimensions only to calculate volume for well control applications.

COILED TUBING CAPACITY

PIPE SIZE OD	WALL THICKNESS	WT/FT	BBL/FOOT
1"	0.087	0.849	0.00069
1"	0.095	0.918	0.00061
1"	0.102	0.978	0.0006
1"	0.125	1.167	0.0005
1 ¼"	0.080	0.9997	0.0012
1 ¼"	0.095	1.172	0.0012
1 ¼"	0.102	1.251	0.0011
1 ¼"	0.116	1.405	0.0010
1 ¼"	0.125	1.502	0.0010
1 ¼"	0.175	2.0082	0.0008
1½"	0.102	1.5233	0.0017
1½"	0.109	1.6234	0.0016
1½"	0.125	1.8355	0.0015
1½"	0.156	2.2391	0.0014
1½"	0.175	2.4764	0.0013
1¾"	0.102	1.7952	0.0023
1 ¾"	0.095	1.6781	0.0024
1 ¾"	0.145	2.4861	0.0021
1 ¾	0.188	3.1363	0.0018
1 ¾	0.203	3.3542	0.0017
2.0"	0.116	2.3341	0.0031
2.0"	0.125	2.5041	0.0030
2.0"	0.134	2.6731	0.0029

Standard Oil Field Tubing Joints Dimension Capacity

Outside diameter	Weight/ft.	Inside diameter	Capacity (bbl./ft.)	Cu. ft./Lin. ft.	
1.900″ (1.610)	2.90	1.61	0.0025	0.0149	67.10′
2.063″ (1.751)	3.25	1.751	0.0030	0.0167	59.80′
2 ⅜″ (2.375)	4.70	1.995”	0.00387	0.02171	46.10′
2 ⅞″ (2.875)	6.50	2.441	0.00578	0.0325	30.77′
2 ⅞″ (2.875)	8.60	2.259	0.00520	0.02783	35.93′
3 ½″ (3.500)	9.30	2.992	0.00870	0.04884	20.48′
3 ½″ (3.500)	10.20	2.992”	0.00830	0.0466	21.47′
4″	9.50	3.43	0.0123	0.0532	10.13′
4″	11.0	3.35	0.0118	0.0511	10.11′
4″	11.60	3.31	0.0114	0.0510	10.14′
4 ½″ (4.500)	12.75	3.598	0.0152	0.0854	11.70′

Cubic foot per linear foot is obtained from the following:

Example: What is the cubic foot per linear foot in 2 3/8″ tubing?

0.00387 divided by 0.1781 = 0.0217

or

0.00387 divided by 5.6 = 0.0217

1 cubic foot in 2 3/8″ tubing is 1 divided by 0.0217 = 46.10′

The Casing Capacities

OD Size	Weight (lb./ft.)	Inside Diam.	Capacity (bbl./ft.)	Cu. ft./ Lin. ft.)	
4 ½″	9.50#	4.09″	0.0162	0.0912	10.96′
4 ½″	10.50#	4.052″	0.0159	0.0895	11.17′
4 ½″	11.60#	4.000″	0.0155	0.0872	11.46′
4 ½″	12.60#	3.958″	0.0152	0.0835	11.83′
4 ½″	13.50#	3.920″	0.0149	0.0832	11.93′
5″	15#	4.408″	0.0189	0.1059	9.44′
5″	18#	4.276″	0.0178	0.0997	10.03′
5 ½″	14#	5.012″	0.0244	0.1387	7.21′
5 ½″	15.50#	4.960″	0.0238	0.i336	7.48′
5 ½″	17#	4.892″	0.0232	0.1305	7.66′
5 ½″	20#	4.778″	0.0222	0.1245	8.03′
7″	17#	6.538″	0.0415	0.2332	4.29′
7″	20#	6.456″	0.0405	0.2273	4.40′
7″	23#	6.366″	0.0393	0.2210	4.52′
7″	26#	6.276″	0.0383	0.2148	4.66′
7″	32#	6.094″	0.0361	0.2025	4.94′
7 ⅝″	26#	6.844″	0.0472	0.2646	3.88′
7 ⅝″	29.70#	6.873″	0.0458	0.2582	3.86′
7 ⅝″	33.70#	6.640″	0.0445	0.2496	4.01′
8 ⅝″	24#	8.097″	0.0637	0.3575	2.80′
8 ⅝″	28#	8.017″	0.0624	0.3505	2.85′
8 ⅝″	32#	7.921″	0.0609	0.3422	2.92′
8 ⅝″	36#	7.825″	0 .0595	0.3340	2.99′
9 ⅝″	29.30#	9.063″	0.0797	0.4479	2.23′
9 ⅝″	32.30#	9.001″	0.0787	0.4418	2.26′
9 ⅝″	36#	8.921″	0.0773	0.4340	2.31′
9 ⅝″	38#	8.885″	0.0766	0.4305	2.32′
9 ⅝″	40#	8.835″	0.0758	0.457	2.35′
9 ⅝″	43.50#	8.755″	0.0745	0.4180	2.39′
9 ⅝″	47#	8.681″	0.0732	0.4110	2.42′

How to use the above ready data:

Example I

What is the capacity of 2,300 feet of 5 ½″ 17# casing?

$$2{,}300' \times 0.0232 = 53.56 \text{ barrels}$$

Example II

What is the capacity of 2,300 feet of 2 ⅞″ 6.50# tubing?

$$2{,}300' \times 0.00578 = 13.29 \text{ barrels}$$

Example III

How much 20/40 sand is needed to fill 100′ in a 5 ½″ 15.50# casing?

0.1336 cu. ft./liner ft. $\times 100 = 13.36$ cu. ft. or 13.36 sacks (100# sacks)

One cubic foot of 20/40 sand is equal to 100 pounds of sand.

100 pounds of 20/40 gravel-packing sand is equal to one cubic foot.

What is the hydrostatic pressure of 8,000 feet of 11 pounds/gallon of calcium chloride water in a 7″ 23# casing?

$HP = 0.052 \times 8{,}000 \times 11 = 4{,}576$ psi

One gallon of fresh water is equal to one pound and equal to 231 cubic feet.

What is the hydrostatic pressure of 8,000′ of 11 pounds/gallon of calcium chloride water in 2″ 4.6# tubing?

$HP = 0.052 \times 8{,}000' \times 11 = 4{,}576$ psi

$BHFP = ISDP + HP$

Bottom hole frac. pressure = Instantaneous shutdown pressure + Hydrostatic head

STP (surface treatment) = BHFP – HP + (FP)

Typical Oil and Gas Wellbore Schematic That Engineers Should Know

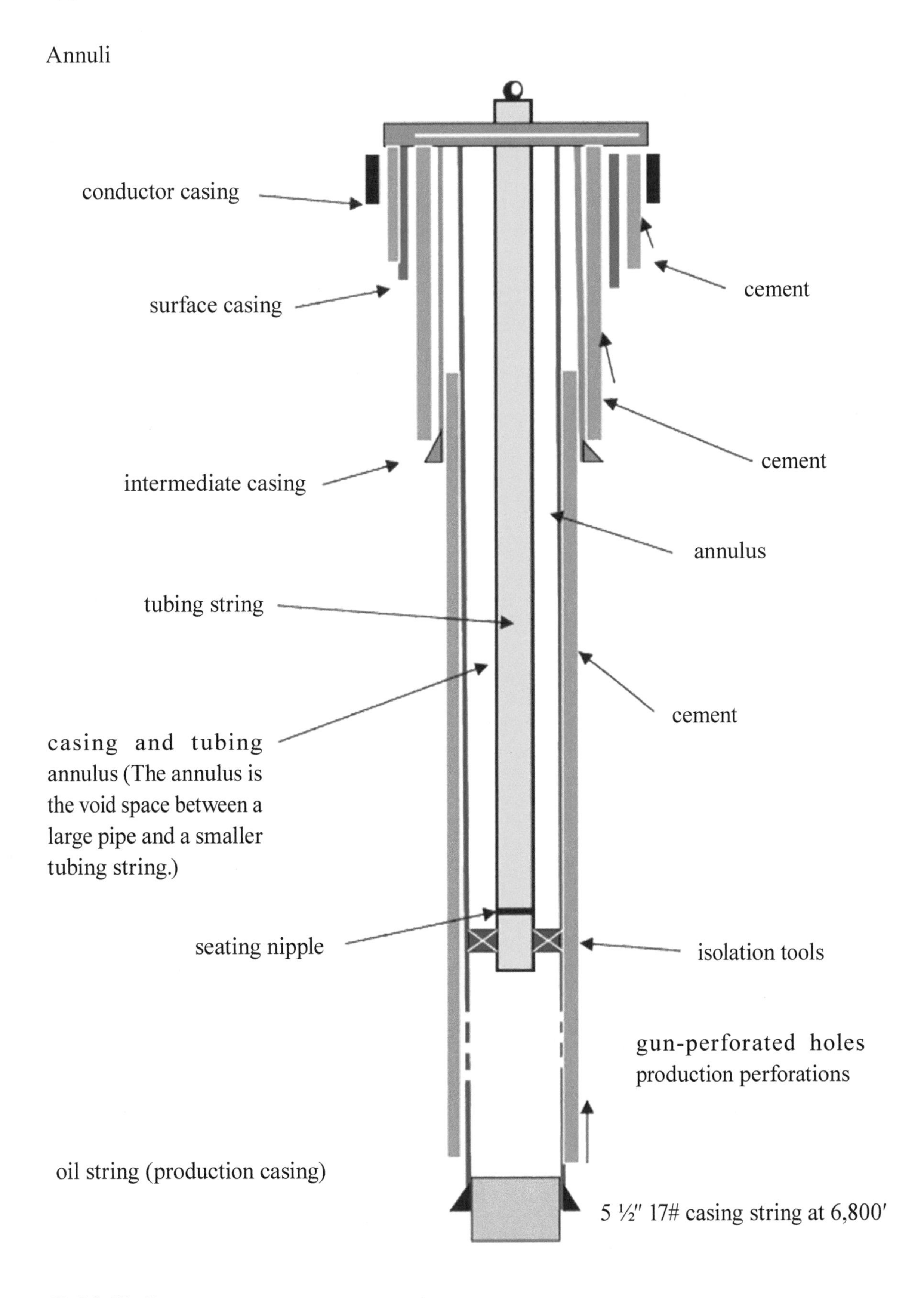

K. M. Hadipour

Volume and Height between the Casing String and Tubing String

The volume and height are basically the capacity of the annulus between a larger pipe and smaller tubing or the volume between a pipe and an open hole. The volume between the outside of the tubing and the inside of a casing is extremely important in gravel packing and wellbore circulation calculations (the casing inside the diameter and the screen and liner outside the diameter).

Example: the annular volume between 5.5″ (15.50#) casing and 2 ⅜″ tubing string

(D = 5.5″; d = 2.375″) $(D^2 - d^2) \times 0.0009714$ (D = inside diameter of the larger pipe; d = outside diameter of the smaller pipe)

$(24.50 - 5.641) \times 0.0009714 = 0.0183196$ barrel per linear foot

Annular capacity between 5.5″ (15.50#) casing and 2 3/8″ tubing string = 0.0183196 barrels

Annulus Volume and Height Capacity (Between Two Strings)

The volume between 2 $^{1}/_{16}$″ (2.0625″) OD tubing and casing strings is shown below:

Casing

Casing Size	Wt. per Ft.	Barrel/Ft.	Cu. Ft./Ft.	Lin. Ft. per Cu. Ft.
4 ½″	9.50#	0.0121	0.0680	14.70′
4 ½″	10.50#	0.0112	0.0663	15.07′
4 ½″	11.60#	0.0114	0.0641	15.61′
4 ½″	13.50#	0.0108	0.0606	16.50′
5″	11.50#	0.0161	0.091	11.09′
5″	13#	0.0155	0.087	11.50′
5″	15#	0.0147	0.0828	12.08′
5″	18#	0.0136	0.0765	13.07′
5 ½″	14#	0.0203	0.1138	8.79′
5 ½″	15.50#	0.0197	0.1104	9.06′
5 ½″	17#	0.0191	0.108	9.32′
5 ½″	20#	0.0180	0.1013	9.87′
5 ½″	23#	0.0171	0.096	10.45′
7″	20#	0.0364	0.2041	4.899′
7″	23#	0.0352	0.199	5.055′
7″	26#	0.0341	0.1916	5.22′
7″	32#	0.0319	0.1793	5.576′

Annular Capacity (Between Two Strings)

The *annular capacity* is defined as "the volume between a casing string and tubing and/or the volume between an open hole and the drill string."

The volume between 2 ⅜″ (2.375) OD tubing and casing strings is listed below:

Example: Seventy barrels of water will fill up how many feet in the annulus between 2 ⅜″ 4.70# tubing and a 5 ½″ 17# casing string?

70 barrels divided by 0.0178 = 3,933 feet

How many barrels of fluid are needed to fill up 7,000′ of annular space between 2 ⅜″ tubing and 5 ½″ 17# casing?

$0.0178 \times 7{,}000' = 124.6$ barrels

Size /Wt. per Ft.	ID Drift	Bbl./Ft.	Cu. Ft./Lin. Ft.	Lin. Ft. /Cu. Ft.
4 ½″ (9.50#)	4.090″	0.0108	0.0605	16.54′
4 ½″ (10.50#)	4.052″	0.0105	0.0588	17.1′
4 ½″ (11.60#)	4.000″	0.0101	0.0565	17.70′
4 ½″ (13.50#)	3.920″	0.0094	0.0530	18.85′
4 ½″ (15.10)	3.826″	0.0087	0.0490	20.38′
5″ (11.50#)	4.560″	0.0147	0.0826	12.10′
5″ (13#)	4.369″	0.0141	0.0794	12.59′
5″ (15#)	4.404″	0.0134	0.0752	13.30′
5″ (18#)	4.277″	0.0123	0.0690	14.51′
5 ½″ (13#)	5.044″	0.0192	0.1080	9.26′
5 ½″ (14#)	5.013″	0.0189	0.1061	9.412′
5 ½″ (15.50#)	4.950″	0.0183	0.1029	9.72′
5 ½″ (17#)	4.893″	0.0178	0.0998	10.024′
5 ½″ (20#)	4.777″	0.0167	0.0937	10.667′
5 ½″ (23#)	4.670″	0.0157	0.0882	11.34′
7″ (17#)	6.536″	0.0360	0.2027	4.94′
7″ (20#)	6.455″	0.0343	0.1925	5.09′
7″ (23#)	6.366″	0.0339	0.1903	5.26′
7″ (26#)	6.276″	0.0326	0.1840	5.44′
7″ (32#)	6.095″	0.0306	0.1717	5.80′

The volume between 2 ⅞″ (2.875″ OD) tubing and casing strings is listed below:

Size	Wt./ Ft.	Bbl./Ft.	Cu. Ft./Lin. Ft.	Lin. Ft./Cu. Ft.
4 ½″	9.50#	0.0082	0.0462	21.67′
4 ½″	10.50#	0.0079	0.445	22.49′
4 ½″	11.60#	0.0075	0.0422	23.71′
4 ½″	13.50#	0.0069	0.0387	25.82′
5″	11.50#	0.0122	0.0683	14.63′
5″	13#	0.0116	0.0651	15.37′
5″	15#	0.0108	0.0609	16.42′
5″	18#	0.0087	0.0546	18.30′
5 ½″	13#	0.0167	0.0937	10.671′
5 ½″	14#	0.0164	0.0651	15.37′
5 ½″	15.50#	0.0158	0.0866	11.29′
5 ½″	17#	0.0152	0.0854	11.703′
7″	17#	0.0335	0.1881	5.32′
7″	20#	0.0325	0.1545	6.47′
7″	23#	0.0313	0.1760	5.68′
7″	26#	0.0302	0.1697	5.89′
7″	32#	0.0280	0.1575	6.35′
7 ⅝″	24#	0.0399	0.2241	4.46′
7 ⅝″	26.40#	0.0391	0.220	4.529′
7 ⅝″	29.70#	0.0379	0.2127	4.70′
7 ⅝″	33.70	0.0364	0.2045	4.90′
7 ⅝″	39#	0.0346	0.1943	5.147′

How many barrels of fluid are needed to fill up 9,000′ of the annular space between 2 ⅞″ tubing and 5 ½″ 17# casing?

9,000′ × 0.01512 = 136.8 bbls. (from the table above)

The volume and height between 3 ½″ (3.50″) OD tubing and casing strings are listed below:

Size	Wt./Ft.	Bbl./Ft.	Cu. Ft./Lin. Ft.	Lin. Ft./Cu. Ft.
7″	17#	0.0296	0.1663	6.02′
7″	20#	0.0286	0.1605	6.23′
7″	23#	0.0275	0.1542	6.48′
7″	26#	0.0264	0.1480	6.76′
7″	29#	0.0252	0.1418	7.05′
7″	32#	0.0242	0.1357	7.38′
7 ⅝″	24#	0.0360	0.2023	4.94′
7 ⅝″	26#	0.0353	0.1981	5.49′
7 ⅝″	29.70#	0.0340	0.1921	5.24′
7 ⅝″	33.70#	0.0326	0.1828	5.47′

Two hundred barrels of water will fill up how many feet of annulus space between 3 ½″ tubing and 7 ⅝″ 26# casing string?

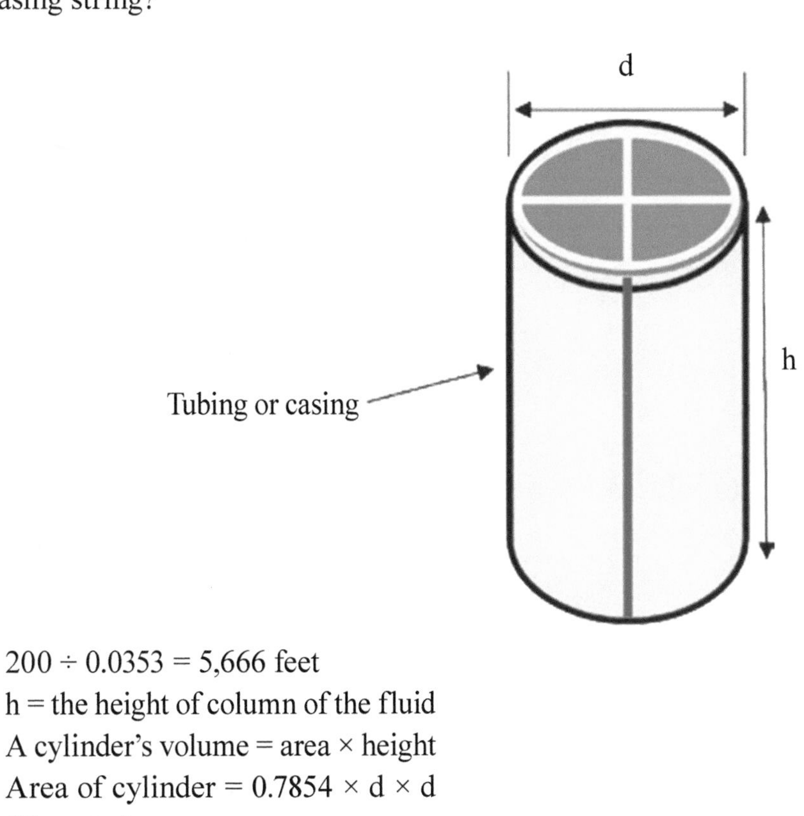

200 ÷ 0.0353 = 5,666 feet
h = the height of column of the fluid
A cylinder's volume = area × height
Area of cylinder = 0.7854 × d × d
(diameter)

pipe

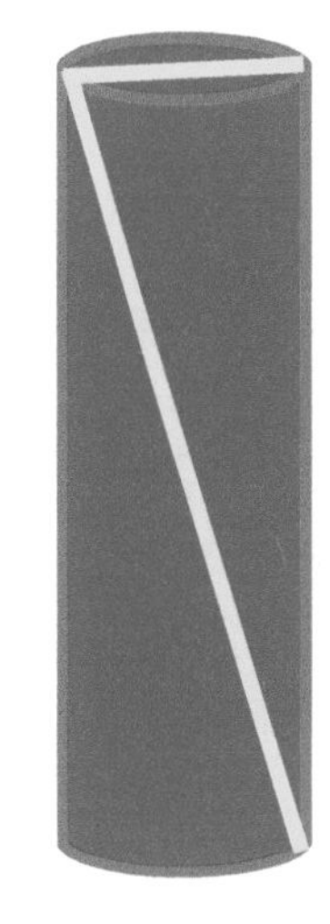

All-Welded Wire-Wrapped Screens

Pipe Outside Diameter		Coupling OD	Screen Outside Diameter
2 1/16″	(2.0625″)	Integral	2.61″
2 3/8″	(2.375″)	2.875″	2.92″
2 7/8″	(2.875″)	3.50″	3.42″
3 ½″	(3.50″)	4.250″	4.05″
4.0″	(4.00″)	4.750″	4.55″
4 ½″	(4.50″)	5.000″	5.05″
5.0″	(5.00″)	5.563″	5.55″
5 ½″	(5.50″)	6.05″	6.05″
7″	(7.00″)	7.656″	7.55″

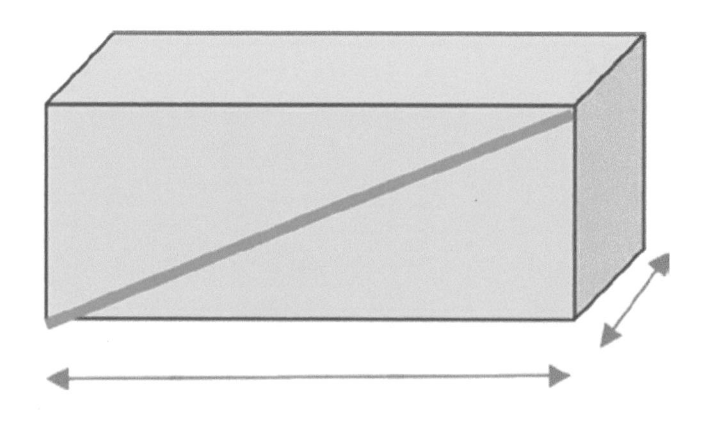

Rig Tanks:
volume = height × width × length
volume = 10′ × 9′ × 18′ = 1,620 cubic feet
1,620 × 0.0178 = 288.5 barrels
1,620 divided by 5.6 = 288.barrels

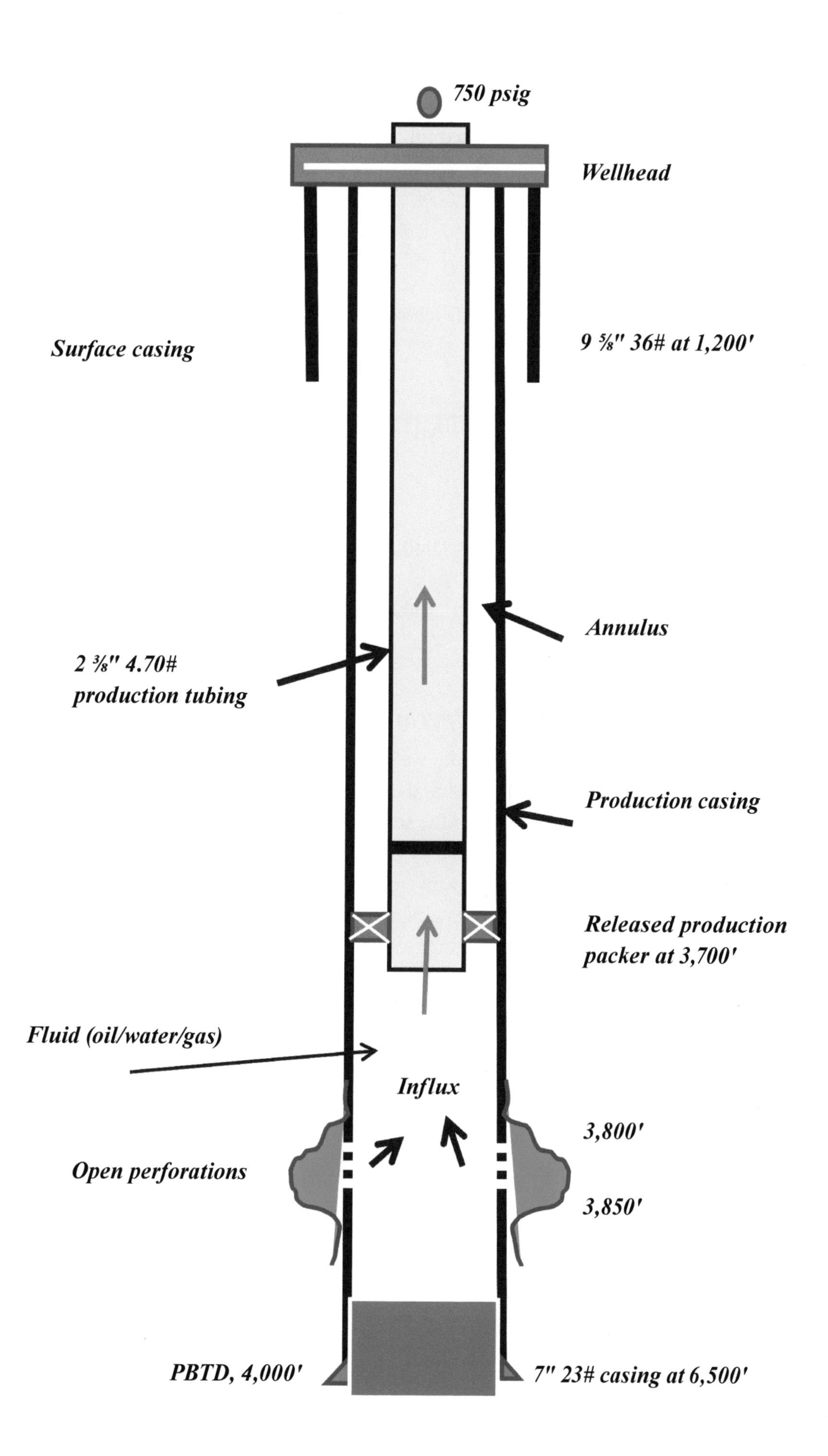
750 psig
Wellhead
Surface casing
9 ⅝" 36# at 1,200'
Annulus
2 ⅜" 4.70#
production tubing
Production casing
Released production
packer at 3,700'
Fluid (oil/water/gas)
Influx
3,800'
Open perforations
3,850'
PBTD, 4,000'
7" 23# casing at 6,500'

Buoyancy Force (Floating Force)

Buoyant force is an upward force from immersed tubing or an object in a fluid. Buoyant force depends on the size of the immersed tubing and fluid density (it is basically referred to as the weight of the liquid displaced).

A) Close-ended pipe:

$$BF\ (lb.) = 0.0408 \times MW \times OD2 \times MD\ (\text{close-ended pipe})$$

B) Open-ended pipe:

$$BF\ (lb.) = 0.0408 \times MW\ (OD2-ID2) \times MD\ (\text{open-ended pipe})$$

Buoyant weight of the pipe:

$$(\text{Pipe weight} \times MD) + (0.040 \times ID2 \times MW) - (0.040 \times OD2 \times MW \times MD)$$

$$BF\ (\text{buoyancy factor using mud weight}) = \frac{65.5 - \text{mud weight (ppg)}}{65.5}$$

Onshore Coiled Tubing Intervention Operation

1. Travel with the coiled tubing unit to the well location.
2. While you are driving on the oil field and encounter narrow dirt roads, bumps, and gravel, slow down to avoid damaging some of the sensitive tools and instruments on the coiled tubing unit.
3. Once you arrive at the location and away from the well, begin analyzing the area.
4. Check the location for power lines, fencing, and other hazard conditions.
5. Do not back up to or rig up on the wrong well.
6. Do not rig up on a wet and soft location.
7. Meet with the company man in charge of the well operations.
8. Demand all the wellbore information and wellbore schematics:
    a. Main objective and purpose of coiled tubing operation to the well
    b. Top connection size on the wellhead (Christmas tree)
    c. Crown valve above the master valve on the wellhead
    d. Wing valve and valves
    e. Wellbore shut-in pressure (read and record)
    f. H_2S and CO_2 content (never assume a well without H_2S gas)
    g. Production tubing size (2″, 2 ⅜″, 2 ⅞″, 3 ½″)
    h. Depth of all the nipples restriction (S-nipple, X-nipple, slide/sleeve with full accurate inside diameter sizes) (make sure all the BHA will go through the nipples and the tight spots)
    i. Packer depth and tail pipes (what is below the packer assembly)
    j. Known objects in the hole across or below the open perforations

k. Casing size and weight (4″, 4 ½″, 5″, 5 ½″, 7″, 8 ⅝″)
l. Perforation depth (top and bottom of open perforations)
m. Total depth to clean up the wellbore (PBTD)
n. Any and all known restrictions in the well (lost or floating fish)
o. Cleanup target depth
p. Downhole tubing and hydraulic controllers (SSCSV, SCSV)

Make sure orders are in writing and clear. Avoid "you said," "he said," and "I said" conflicts during or after the job.

Water Disposal, Injection Wellbore, Low-Pressure Gas Well Cleaning
Over-the-Land Operations

Some oil and gas well locations are small with narrow roads, wet ditches, and dangerously slick and soft shoulders across oil fields (equipment may get stuck on sandy roads or even on the well location because of the heavy coiled tubing unit).

Often it is difficult to back up the equipment into a location to line up the coiled tubing unit. You may use a winch truck to assist you in case of emergency. Most modern coiled tubing units are equipped with electronic control equipment that must be protected. Shaking, jerking, and vibration may cause damage to sensitive computer devices and may fail in the middle of coiled tubing tripping operations going in or out of the well.

Preparing to Rig Up Coiled Tubing on a Well

1. Move in and spot the coiled tubing unit at the well (use two spotters to back up the unit to the well to avoid running into the wellhead). Rig up on a solid hard ground area at the well.
2. Spot nitrogen unit close or parallel to the coiled tubing unit.
3. Spot crane (high-reach crane is useful in tight well areas).
4. Spot fluid pump if it is a separate unit.
5. Spot a circulating tank (steel pit) for backwashing the solids from the well. The reverse pit or fluid tank may be located 60′ away from the wellbore if you are working on an injection or saltwater disposal well.
6. Spot a water supply truck (supply of clean produced water for coiled tubing). Do not use fresh water in washing, circulating, and wellbore control.
7. Spot empty vacuum truck (extra vacuum truck to haul dirty water).
8. Conduct a safety meeting (discuss operation steps and safety measures).
9. All the workers must wear their PPEs: hard hats, steel-toed boots, safety glasses, clean long-sleeved shirts and pants, and working gloves.
10. Spot fire extinguishers at a reach.
11. Spot H2S monitor equipment (never assume a well is without H2S).
12. Point out the nearest hospital in case of an emergency (who will drive?).
13. No smoking within 150′ of the wellhead.
14. Check and monitor H2S present.
15. Once all the equipment is spotted at the well location, open the well and read and record shut-in tubing and shut-in casing pressure. (Some saltwater disposal wells may flow back with high flowing pressure.)
16. Evaluate the wellbore conditions. Is the coiled tubing and pressure control equipment sufficient to handle any unexpected wellbore pressure before and after the wellbore intervention?

17. If the casing pressure (annulus) is high (abnormal) and the production tubing integrity is questionable, do not rig up on the well (possible packer leak, hole in tubing, parted tubing string, and/or hole in the casing string). Stay out of trouble.
18. Let the company man in charge of the operation and your boss know about all your abnormal findings and concerns.

[See wellbore schematic.]

19. For any abnormal well conditions, you must consult with your supervisor and the company man present at the well site.
20. If the casing pressure is not questionable, continue rigging up on the well.
21. Check the Christmas tree (make sure the wellhead is stable).

    The Christmas tree is the major segment of the wellhead above the ground. The Christmas tree is flanged up and fastened on top of the wellhead to control well flowing pressures.

    Christmas trees and wellheads may appear in different shapes, sizes and configurations, consisting of two master gate valves (lower-master and upper-master valves), one crown valve (swab valve), and one wing valve.

    Some Christmas trees are tall and huge. Some Christmas trees and production tubing strings are equipped with emergency safety valves that must be opened and closed properly to avoid downhole damages. There are two types of control safety valves in oil and gas well operations (offshore and onshore):

    - Surface control safety valves (SCSV)
    - Subsurface control safety valves (SSCSV)

The SCSV and SSCSV are hydraulic pressure–controlled valves that must be opened or closed by the company man in charge of the operation to avoid accidental valve closure during the coiled tubing operation.

For on-the-land operations, most low-pressure wells do not have what I call standard normal high-pressure–rated Christmas trees. Some companies normally get away with anything in the name of cost-cutting.

Most of the land wells are completed with single low-pressure wellheads such as the Larkin head type, which is welded or screwed onto the top of the production casing string (you will face screw-type caps with slips and rubber pack-offs). The slip assembly in the Larkin-type wellhead assembly is designed to hold the production tubing from falling or to prevent the pipe from slipping down the hole. It is not designed to hold the tubing from upward movements.

During high injection pressure or circulating pressure, the tubing may move upward (packer is subject to be forced up the hole). Most of the low-pressure artificial lifting wells such as electric submersible wells, pumping wells, PCP pumping wells, saltwater injection wells, and saltwater disposal wells may only have one low-pressure master valve and one wing valve to operate with (poor boy hookup).

22. Before you start working on any well for coiled tubing interventions, demand to install a reliable full opening crown valve above their master valve (swab valve above the master valve for rigging up the coiled tubing well control stack).
23. I prefer flanged-up connections on all the oil- and gas-producing wells rather than threaded connections before going into a well.
24. Install blowout preventer stack. Check and function-test all the rams. Replace any used pack-off element.
25. Never shortcut on any safety equipment to save a few dollars. It will catch up with you (never underestimate wellbore pressure).

Rigging Up Coiled Tubing on Saltwater and Injection Wells

26. Make sure the hydraulic-operated coiled tubing reel slowly spools-off the coiled pipe from the service reel and is pushed through the level wind and coiled tubing's depth meter (zero the depth counter).
27. The tube will be forced over the gooseneck arch assembly over a series of steel rollers and dies (tubing guides).
28. The lubricated tube will continue rolling/traveling from the gooseneck arch through the injector head and out of the stripper pack-off assembly.
29. The coiled tubing end will be checked for sharp bends, dents, and bad spots.

A few feet of the coiled tubing may be cut by the operator to have a good-condition tubing end without an oval shape, bends, or dents - to be used for the bottom-hole assembly (most coiled tubing operators know the history of coiled tubing equipment). The used mule shoes or jet nozzle will always be cut off and will be replaced with new bottom-hole assembly as required.

30. The selection of bottom-hole assembly is based on the type of coiled tubing work.
31. Connect and install the nitrogen line and water fluid lines to the reel manifold.
32. Pick up on the coiled tubing assembly and install new bottom-hole assembly such as jet nozzle, mud motor, or plain mule shoe as requested (test connection to make sure it is in satisfactory operating condition).
33. The bottom-hole assembly connection to the coiled tubing string must be leak proof and reliable (should be as strong as the coiled tubing itself).
34. It is a good professional practice to fill up and test the coiled tubing for leaks after each wellbore cleaning and/or acidizing (either in the yard or at the well site).

    Check the physical condition of the Christmas tree or wellhead above the surface ground. The Christmas tree is the top connection above the wellhead with several valves (master valves, wing valves, crown valves, and casing valves). Always install a full opening high-pressure crown valve above the master valve to prevent you from closing the master valve for any reason prior to rigging up on the well.

    The coiled tubing operator may choose to cut the end of the coiled tubing like a mule shoe and/or install a new jet nozzle at the end of the coiled tubing string. Choosing a jet nozzle or a simple mule shoe is based on the coiled tubing operator and/or the customer representative to select the bottom-hole assembly (depends on the downhole operation).

35. I prefer to circulate kill fluid and retest the coiled tubing prior to going into the well because of several past experiences.

    - I do not want you losing the pipe into the well due to a hole in the tubing.
    - Testing the coiled tubing will give you a comfort of mind and trouble-free operation before tripping into the wellbore.
    - You will not be able to wash solids with a hole in the coiled tubing string, and it may cost you trip time and possible fishing work.
    - You will not be able to circulate fluids to kill a well with a hole in the coiled tubing string.
    - Fill up the coiled tubing with water and test the coiled pipe (always conduct an integrity test as required) to detect tubing leaks before going into the hole (using a pressure recorder is preferred).
    - This procedure may detect and prevent the coiled pipe from becoming parted in the well while washing and circulating with high-pressure nitrified gas below the wellhead.
    - If you are using a drill bit or bladed mill on a mud motor, pull and function-test the bottom-hole assembly. Never run any questionable tools and equipment into a wellbore to cut cost.

Selection of the Bottom-Hole Assembly

(It is based on what you are planning to do.)

36. There is no such thing as simple coiled tubing work (anything can go wrong during a coiled tubing operation).
    Higher annular fluid velocity is necessary to move solids in solution to prevent sand bridging. Circulation rate and pressure is a critical factor when washing solids inside large casing and/or production strings. Adding viscous liquid gel to the circulating fluid will suspend solids during circulation and may prevent the coiled tubing from becoming stuck! Application of N2 will reduce hydrostatic head, and accelerate the rate of circulation. The purpose of gel polymer is to provide sufficient viscosity in fluid to suspend solids during well circulation in the form of slurry and/or slugs. The selected gels will normally break-down with time and temperature.

37. The bottom-hole assembly on coiled tubing operations may consist of various types of tools based on the work objectives of the well intervention.

    There are several choices of bottom-hole selections:

    - Mud motor assembly:
      - ✓ Rock bit or bladed mill
      - ✓ Bit sub
      - ✓ Mud motor
      - ✓ A few drill collars
      - ✓ Crossover sub
      - ✓ Releasing tool
      - ✓ Check valves
      - ✓ Connector sub back to the coiled tubing string

- Jet nozzle assembly

- ✓ Pointed jet nozzle with holes
- ✓ Check valve
- ✓ Coiled connector

DD

- Mule shoe assembly

  If you select to cut the coiled tubing like a mule shoe, you may not have a check valve on the coiled tubing string. The application of a mule shoe on a coiled tubing operation is only good for cleanup operations in saltwater disposal/or injection wells.

  Mule shoe shape on the coiled tubing is just cutting the coiled tubing at the end in a slanted angle. (Do not use sharp-pointed mule shoes.) I prefer to cut a dual mule shoe without sharp-pointed tips.

- Pick up the injector a few feet off the ground to install the bottom-hole assembly.

  If you plan to use a jet nozzle, the end of the coiled pipe must be freshly cut and smooth and internally cleaned up using a rotary grinder to de-bure.

- The jet nozzle tool will be forced into the coiled tubing and crimped tightly, as strong as the coiled pipe itself (the connector must not leak and or come off).
- More often, the connection to the coiled tubing will be pull-tested to ensure the bottom-hole assembly is securely in place.

  Installing the type of wash tools are based on the coiled tubing project requirements (may use a mud motor with bits/mill, jet nozzle and/or plain mule shoe)

- If the bottom-hole assembly is too long, the wellhead will be spaced with larger-diameter tubing pup joints (riser) to cover the entire bottom-hole assembly length.
- Coiled tubing spacing and rig-up configurations often become very complicated and risky.
- The riser spacing is a practical safety measure when pulling the tool up the hole under pressure (acting as a lubricator).
- The spacing riser will be located below the injector assembly and the stripper packing off to the lower valve.

Rigging Up the Equipment on the Well

- Install full opening and high-pressure–rated crown valve above the top master valve (function-test the crown valve to make sure it opens and closes properly and is in good operating condition). (Flanged gate valves are preferred above the master valve for high-pressure coiled tubing operations.) The application of a full opening crown valve above the master valve is a good and safe practice to avoid shutting the master valve on the well during and/or after operation.

- Pick up and install BOP stack on the crown valve (install hydraulic lines and cables as you wish and function-test or repair rams if necessary).
- Rig up and install the riser if necessary (high-pressure wing union-type sub connections are preferred).
- Install and level off the injector over the blowout preventers on the riser.
- Pick up the entire assembly and prepare to check and test the bottom-hole assembly.
- Pick up the entire assembly (BHA and the injector assembly) and make up on the BOP stack.
- Install flow line for fluid return from the well to the steel tank with a gas buster.
- Chain down and secure the flow lines to avoid shaking and vibration on the line during the fluid surge out of the well (a high-pressure surge may break the line in the middle of the coiled tubing operation).
- Use high-pressure wing union-type subs and swivel joints for fluid discharge. Avoid using threaded tees, elbows and ninety degrees as much as possible (do not use high-pressure rubber hoses in the coiled tubing operation). Tees and ninety-degree connections may become cut by fluids in the middle of the operation and force you to shut down.

PEELER RANCH

- Measure the spacing above the master valve if necessary (spacing must be longer than the bottom-hole assembly).
- Flange up the BOP stack on the wellhead and make up the coiled tubing quick connection on the wellhead assembly while holding tension on the coiled equipment to minimize any weight on the wellhead (replace O-ring on Acme quick connection to avoid leaks and shut down during coiled tubing operations under pressure).

1. Finish rigging up the coiled tubing unit and all the supporting equipment.
2. Install high-pressure steel lines from the nitrogen truck to the coiled tubing manifold.
3. Install high-pressure steel lines from the fluid pump to the coiled tubing manifold.
4. Fill up the coiled tubing with (nontoxic) circulating fluid. Provide MSDS sheet of any chemical used in the fluid circulation (agents, acid, others).
5. Check and test equipment for leaks. Repair leaks before going into the well.
6. Check, test, and repair any and all the pressure control elements
7. Check Acme connection and O-rings to avoid leaks during operation.
8. Check all hydraulic systems inside and outside of the control box.
9. Check and zero the coiled tubing depth meter (may add elevation depth).
10. Make sure the depth meter is accurate enough before tripping into the well. Some meters have been found to be off by 100′ (100′ of mistakes off the target depth).
11. Lubricate the coiled tubing to minimize friction and drag going through the injector and pack-off elements.

$$\text{Capacity of Coiled Tubing on the Reel } = \frac{L \times \left[D - (2T)\right]^2}{1029.4}$$

Ready to Trip in the Hole with Coiled Tubing

12. Check all equipment to make sure they are in order before starting the job (coiled tubing tools, fluid pump, nitrogen supply, sufficient water, and vacuum trucks, including coiled tubing fueling).
13. Know the coiled tubing capacity and displacement volume in case of emergency.
14. Have a safety meeting with everyone at the location (define work objective).
15. Everyone should have a duty in case of emergency.
16. Fill up the coiled tubing (on the reel) with water (kill fluid).
17. Test stripper packing and rams of high and low pressure (as required).
18. Open well; read and record shut-in well pressure before entering the well.
19. Bleed off well pressure if necessary before entering into the well.
20. Start going in the hole while filling up coiled tubing with fluid (nitrified water), checking for leaks on coiled tubing and all the equipment above the wellhead.
21. Stop 300–500′ below the wellhead to obtain full circulation at the surface (while filling up the hole). You may comingle the liquid with low-volume nitrogen (nitrified fluid).

If the well flowing back too much water, it may not be necessary to use nitrogen to comingle with the saltwater. Nitrogen will reduce the hydrostatic head of the well fluid and accelerate the flow faster up the hole (no need to waste nitrogen).

22. Some saltwater disposal wells will have considerable shut-in pressure, indicating that the well is charged up and may flow back with high-rate saltwater and some formation sand and solids (you may see oil with traces of natural gas in the solution).
23. You may not need to supply too much water to the location, to wash a well while the well is flowing water during the coiled tubing operation.
    a. You may need to haul off excess washed water during the coiled tubing operation to prevent saltwater and oil spill on the location.

       You may need to keep one or two vacuum trucks to haul water and prevent you from shutting down in the middle of the operation.

    b. Often it may be difficult to catch up with the flowing backwash water on some wells (you may not have enough room to store the water in order for the vacuum truck to catch up).

24. Continue going in the hole while jetting and washing down the wellbore. Check coiled tubing while spooling off, looking for leaks on the wellhead and holes in the coiled tubing.

    Check the annulus for pressure communications. Report abnormal annulus pressure (annulus is the space between the production tubing and outside of coiled tubing string) In a coiled tubing operation, we circulate fluids the long way only. Circulating fluids down the coiled tubing and out of the production tubing annulus is the conventional method.

    Obtain a good continuous circulation before going down the hole. Monitor well circulations at all times (if you are using nitrogen in the fluid, you may expect circulation delays or intermittent circulations). Check the circulation for abnormal well flow at all times.

25. Continue going down through the production tubing and/or the casing string at a moderate speed of 60–70′ per minute while pumping and circulating down the coiled tubing and out of the annulus. Monitor well pressures to detect early abnormal wellbore influxes into the wellbore.

26. In a coiled tubing operation, you cannot reverse the circulation. Your pipe will have a one-way circulating check. You are only able to circulate the long way (circulating down the coiled tubing and out of the annulus between the outside of the coiled pipe and the inside of the production tubing or the casing string, whichever it may be).

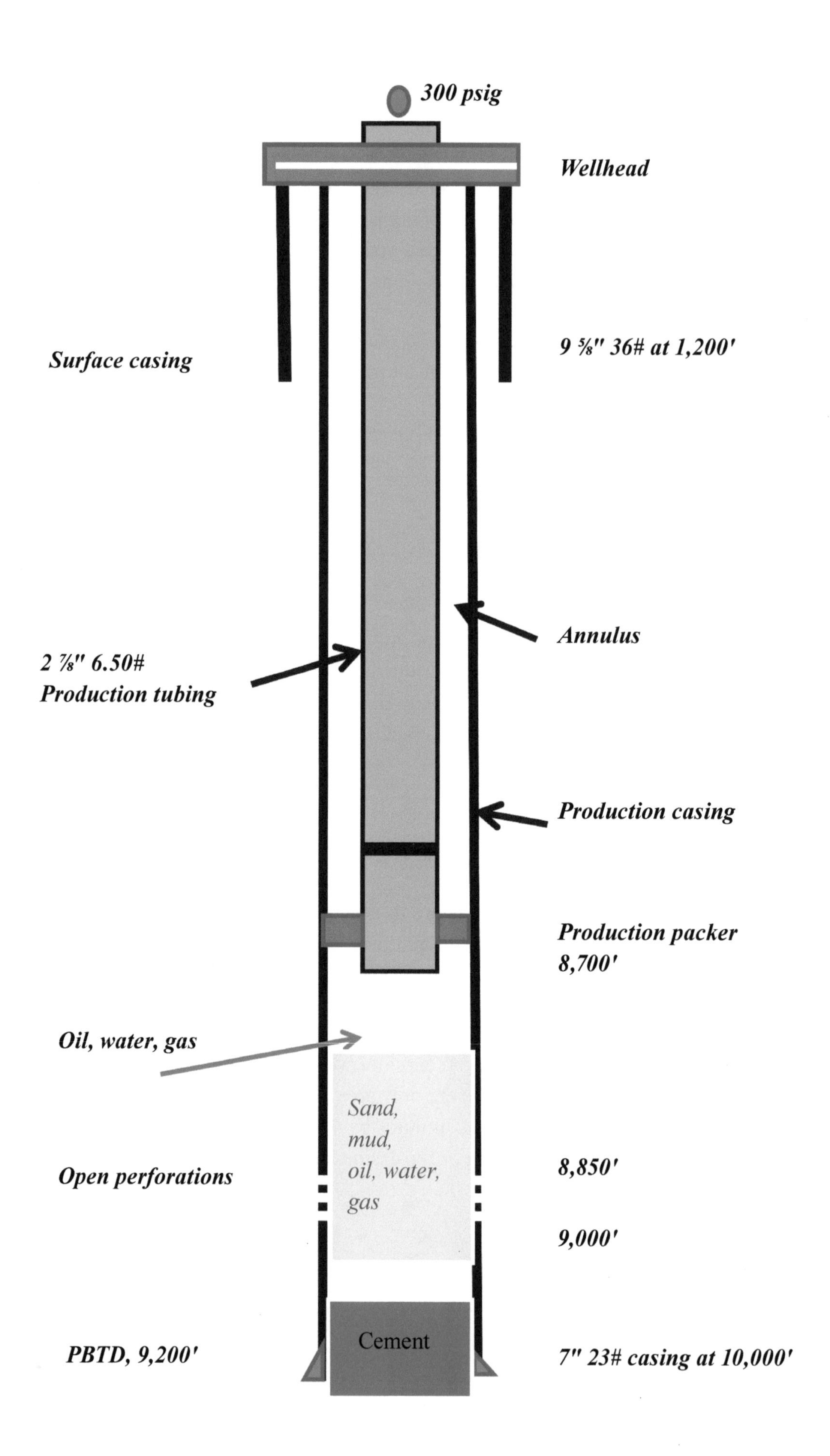
300 psig
Wellhead
Surface casing
9 ⅝" 36# at 1,200'
Annulus
2 ⅞" 6.50#
Production tubing
Production casing
Production packer
8,700'
Oil, water, gas
Sand,
mud,
oil, water,
gas
Open perforations
8,850'
9,000'
Cement
PBTD, 9,200'
7" 23# casing at 10,000'

27. The annulus must be kept fully open without restrictions in order to successfully wash and circulate the bore hole with kill fluid.
28. Catch a fluid sample at the surface and keep a good record of circulated material coming out of the wellbore (sand, mud, shale, oil, gas, or rocks).
29. Continue washing and circulating fluids going into the well.
30. Stop and pick up to check weight every 500′ or as necessary before tagging up on the sand and solids obstruction. Check the pipe to make sure the coiled tubing is free with normal weight. Keep a good record and notes on a job log.
31. When tagging solids or restrictions, stop going down. Pick up and check weights while continue circulating the bore hole. Make sure you have a good fluid circulation before washing and displacing solids.
32 Continue wash down with coiled tubing. Wash solids slowly, not more than 10′ per minutes, while allowing enough time for the fluid to break down and wash the solids apart to avoid bridging up around the coiled tubing.
33. Washing solids while going down must be carried out at a slow speed, not more than ten feet per minute. We are not in the racing business; safety and job quality always come first.

    Work and pick up the pipe off the bottom after every 50–100′ to check for weight, drag, and restrictions while circulating with a good continuous fluid rate. Never stop the pipe for too long when washing and circulating sand, shale, and rocks. (Often you may see large size formation rocks coming out with dirty fluid). The pipe must be in constant move, up or down motion, in order to avoid getting stuck or cause bridging (stay at constant alert while washing or drilling solids). Monitor abnormal pressures trapped below the sand and solids plugs!

    Avoid jamming and running the bottom-hole assembly into the hard solids. You may plug off the bottom hole assembly, bend, and cause kinks on the pipe while working in the casing string. If you are washing with a 1 ¼″ pipe inside a 7″ casing, you should have enough imagination of the sand volume that is packed in the large size casing that must be washed out slowly and carefully.

34. Wash and circulate the well clean to plug back depth (PBTD or as recommended by the well operator).
35. Work the tubing up and down at a slow speed across the open perforation while circulating and keeping the solids sample at the surface.
36. Once the wellbore fluid becomes clean, you may pump the gel sweep down the coiled string (if necessary) while continue washing and circulating with water (Do Not Stop Circulating).
37. Continue to circulate the wellbore until a gel pill or foam is circulated out of the wellbore. Cut the nitrogen off and continue circulating with plain water until all the nitrogen is out of the wellbore. After all the nitrogen is circulated out, start pulling the coiled tubing out of the hole slowly while circulating with water (the well may or may not go on vacuum due to hydrostatic head of water).

38. Pull the coiled tubing above the perforations while still circulating with water.
39. Shut down the return line at the surface while still pumping through the coiled tubing (bull-heading clean water into the well).
40. Start injecting saltwater using a saltwater disposal pump into the well while pulling the pipe and pumping through the coiled tubing string to avoid solids into the coiled pipe. Monitor the injection rate through open perforations. Do not leave the bottom until the well takes an acceptable volume of injecting fluid.
41. Continue pulling the coiled tubing out of the hole slowly, not faster than 50′ per minute, while pumping and injecting water at the same time through the coiled tubing and through the production tubing string (bull-heading water through the coiled tubing and the production tubing at the same time).
42. When the coiled tubing approaches the wellhead at the surface, shut down the coiled tubing fluid pump while injecting field saltwater into the well.
43. Slowly pull the coiled tubing in the clear above the crown valve without interrupting the water injection into the well.
44. Check the coiled pipe coming through the crown valve. Once the pipe reaches above the crown valve, close the crown valve only.
45. With the bottom-hole assembly above the crown valve, you may then displace the water in the coiled tubing with nitrogen into the rig tank (some coiled tubing strings may hold more than 20 barrels of water).
46. Rig down the coiled tubing unit and clean up the location.
47. Install wellhead lines, meters, and pressure gauges; move off the well.
48. Remove wash water and clean up the rig tank (tank may contain mud, sand, silt, traces of oil, and nasty black iron sulfide water).
49. All the circulated material must be hauled off and disposed properly.
50. Obey confined space laws. No one is allowed inside a tank deeper than 4′ without proper gear.
51. Clean up the area and turn the well over to the operator (record and report the injection rate and the injection pressure)
52. Make an impression by doing safe and high-quality work.

***

Application of Coiled Tubing in Oil and Gas Wellbore Interventions

Working on oil and gas wells offshore and/or onshore requires a trained and experienced coiled tubing crew as well as the best available equipment. Some oil and gas wells will have low formation pressure, and some wells may have significantly higher pressure after the remedial cleanup operation.

The coiled tubing wellbore intervention operation is carried out similar to the open-hole conventional drilling and circulating method. It means you will circulate down the coiled tubing string and out of the annulus.

A reverse circulate washing of sand and solids is not a practical procedure in a coiled tubing intervention operation.

- Small coiled tubing string will be plugged off with sand and solids quickly.
- The check valve on the coiled tubing will not allow you to reverse-circulate through the pipe.
- The coiled tubing ID (inside diameter) must be kept fully open and filled with clean kill fluid at all times.
- The coiled tubing must stay internally clean and must be prepared for any downhole well control scenario.

Workover rigs using tubing or drill strings will have the choice to circulate either the long way (pumping down the tubing string and out of the annulus) or circulate the short way (pumping down the annulus and out of the tubing string). On the workover rigs, circulation to kill a well is conducted by pumping kill fluid with either the reverse-circulating method or the conventional method.

Reverse-circulating fluids in workover operations works very well for well control and/or washing solids. Reverse circulation in workover operations will save time compared to the conventional method of pumping down the tubing and out of the casing (it is time-consuming to get the bottoms up). Normally, there is no inside BOP (IBOP) or check valve in the work string tubing during remedial workover operations.

***

There are two major reasons why you should not reverse-circulate fluids during a coiled tubing operation:

a. Coiled tubing is not designed to reverse-circulate. The coiled pipe will have an internal check that will not allow you to reverse-circulate (the purpose of the check valve in the coiled tubing is to keep the coiled pipe full of uncontaminated virgin kill fluid at all times and to avoid the high-pressure influx of oil and gas from entering the coiled tubing).
b. There is no practical purpose for reverse-circulating in a coiled tubing operation. (Simply, it will not work and may complicate the task.) ***Small size coiled*** tubing will be plugged off with sand and will be impossible to clean ***or to remove sand plug out of the pipe*** You may be able to bullhead kill fluid down the annulus and into the open perforations if necessary (as recommended).

Circulate the kill fluid while maintaining the hydrostatic head slightly over the reservoir formation pressure to avoid unwanted kicks (influx into the wellbore). *Influx* is an oil field term for high-pressure reservoir fluid entering the wellbore. Higher pressure–producing wells will allow crude oil, natural gas, and solids to flow under considerable surface pressure.

HP
The fluid hydrostatic pressure is very important pressure in the wellbore control. HP is the pressure created by the height of column of fluid in the wellbore! HP depends on the applied fluid gradient (weight/density).

The wellbore pressure control: controlling the well pressure is carried out either by heavier fluid and/or mechanical barriers such as valves and/or chokes!

To calculate the hydrostatic pressure in a wellbore with multiple fluid densities it is necessary to calculate the sum of hydrostatic pressures across each interval.

***

Coiled Tubing Work on an Oil and Gas Well

Conducting a coiled tubing operation in oil and gas well intervention is carried out with careful considerations, trained and experienced personnel, dependable tools and equipment with capabilities to complete the job safely and cost-effectively.

On normal oil or gas intervention operations, you may need the following tools and equipment to carry on with the job safely and efficiently:

- Trained, knowledgeable, and experienced coiled tubing operator and crew
- Work procedure and planned guidelines
- Defined plan and work procedure (everyone needs to sign)
- Good-standing coiled tubing unit
- Good-standing coiled tubing string
- Nitrogen unit (if needed)
- Good-standing fluid pump
- Vacuum truck/trucks
- A high-reach crane or cherry picker
- Open top steel tanks or frac tanks
- Updated and dependable pressure control equipment
- Newly installed and good-quality BOP rams
- Kill sheet preparation before starting

Transport the equipment to the well location and meet with the company man in charge of the well operations. Demand all the wellbore information, including but not limited to the following:

a. Purpose of working on the well using a coiled tubing operation
b. Top connection size on the wellhead (swab valve on top of the Christmas tree)
c. Full opening crown valve above the master valve on the well
d. Wing valve and check valves (remove checks on return line)
e. Wellbore shut-in pressure (read and record)
f. H_2S and CO_2 content (never assume a well without H_2S gas)
g. Production string size and weight (2 ⅜", 2 ⅞", 3 ½", 4 ½")
h. Coiled tubing size, ID, OD, capacity, and tensile strength
i. Depth of all nipples restriction (S-nipple, X-nipple, slide/sleeve with full accurate inside diameter sizes)

j. Hydraulic actuated surface and subsurface valves (SCSSV and SSCV) (make sure all the BHA will go through the nipples and the tight spots)
k. Packer depth and tailpipes (what is below the packer assembly)
l. Any known screen and liner assembly (top and bottom depth)
m. Known objects in the hole across or below the open perforations
n. Casing size and weight (4″, 4 ½″, 5″, 5 ½″, 7″, 8 ⅝″)
o. Perforations depth (top and bottom of open perforations)
p. Total depth to clean up the wellbore (PBTD)
q. Any and all known restrictions in the well (lost fish)
r. Target cleanup depth (make sure orders are in writing and clear)
s. Well fluid oil, gas, water, mud (you need the fluid weight for calculation)
t. Maximum anticipated wellbore pressure
u. Who is going to do what in case of an emergency

Various Kill Fluid Densities

Fresh water	8.3 ppg
Field salt water	8.6 to 9 ppg
Sea water	8.4 to 8.5 ppg
Brine water	8.4 to 9.9 ppg
KCL water	8.4 to 9.6 ppg
CACL2	11.0 to 11.5 ppg
CABR2	11.8 to 15 pph
Drilling mud	10 to 18 ppg
ZBR2	15 to 20 ppg

Avoid order conflicts of "you said," "he said," and "I said" on the job.

Check the surface well condition and the Christmas tree integrity. Let the company man check and function all the valves on the Christmas tree (Surface Safety and Subsurface Safety Valves).

Check any pneumatic controlled valve to ensure function properly.

Christmas trees, and wellheads may appear in different shape, size and configurations consists of one or two master gate valves (lower master and upper master valve), one crown valve (swab valve), and one wing valve (some Christmas trees are tall and huge).

Some Christmas trees and production tubing string are equip with emergency hydraulic pressure safety valves that must be opened and closed properly to avoid downhole damages.

There are basically two types of hydraulic actuated control safety valves in the oil and gas well operations (on offshore and onshore wells)

- Surface control safety valves (SCSV)
- Sub-surface control safety valves(SSCSV)

The SCSV and SSCSV are hydraulic pressure controlled valves that must be opened or closed by the company man in charge of the operation to avoid accidental valve closure during coiled tubing operation!

Install a new high-pressure and full opening flanged-type crown valve on top of the master valve or hydraulic pneumatic controlled valve. I prefer all flanged connections above the ground to be tested high and low.

Avoid using threaded connections as much as possible. Read and record the well's shut-in tubing and casing pressures. If there is an isolation packer in the well, check the annulus to make sure the packer is holding and there are no abnormal communications in annulus.

High pressure in the annulus is an indication of the following:

- Packer leak
- Tubing leak or parted tubing string
- Casing leak

Coiled tubing operations inside of a parted tubing string or with holes in the tubing may cause the coiled tubing to become stuck or parted. Spot and rig up the crane/cherry picker assembly for maximum reach over the wellhead. Flange up and install the BOP stack on the Christmas tree above the crow valve (swab valve).

Space out and install the riser assembly if necessary (the riser assembly length depends on the length of the bottom-hole assembly). Use high-pressure wing union-type subs and swivel joints for fluid discharge. Avoid using threaded tees, elbows and ninety-degree connections as much as possible. Tees and ninety-degree connections will tend to cut with fluids during the wellbore intervention and may force you to shut down in the middle of a sensitive operation.

You must wear safety belts if working 5′ or higher above the wellhead floor. Pick up and install the injector assembly above the wellhead. Install and test the bottom-hole assembly. The bottom-hole assembly in coiled tubing operations may consist of various types of tools based on the types of well intervention.

- Mud motor assembly
    a. Connector sub from the bottom-hole assembly to the coiled pipe
    b. Positive tandem check valve assembly
    c. Releasing tool (disconnect in case of emergency)
    d. Crossover sub
    e. Joints of drill collars (stiff connection)
    f. High rpm mud motor
    g. Bit sub
    h. Rock bit and/or three-bladed mill

- If the selected bottom0hole assembly is a jet nozzle (pointed jet nozzle with new open holes), check the valve and coiled connector. The application of a jet nozzle is only effective in washing and circulating soft formation mud and sand out of the wellbore. Some formation deposited elements are difficult to remove using a jet nozzle alone or mule shoe tips.
- Application of mule shoe: I do not recommend using a mule shoe in oil- and/or gas-producing wells for pressure safety purposes.

- The application of the jet nozzle wash tool and mud motors is preferred in oil and gas cleanup operations.
- Finish rigging up on the well. Check and pressure-test all the well control equipment components.
- Check the wellhead pressure before snubbing through the injector head. If the pressure is too high (over 1,000 psi), you may consider bleeding the well pressure off and/or killing the well to be able to get into the well.

Light coiled tubing is subject to be forced out of the well because of high wellbore pressure. This may occur when a small length of empty light pipe enters a high-pressure wellbore condition. The pressure may force the coiled pipe, like a piece of straw, out of the injector and head toward the coiled tubing reel, creating a large pipe loop over the injector head and the gooseneck (the sight is unpleasant and unsafe).

A similar scenario may occur when running or pulling a packer out of the casing string in shallow depth under high reservoir pressure (you will not have enough of a defense to well control). You may consider killing a high-pressure oil and gas well before rigging up coiled tubing for intervention purposes.

Fill up the coiled pipe with kill fluid before tripping in the hole to prevent a possible influx of gas or oil seepage through the check. Stop the coiled tubing at 500′ while circulating the kill fluid. Check full fluid circulation. Wait for the full circulation through the coiled tubing and out of the annulus using the kill fluid before continuing into the well. When fluid circulates out into the rig tanks, check all the connections at the tree, surface tools, and equipment for dripping and/or leaking (small leaks can become a huge problem).

Continue rolling and sliding the coiled pipe down the hole at a slow speed of 50–70′ per minute while washing and circulating the wellbore with kill fluid. Check the annulus for any abnormal pressure gain/loss. Continue into the hole. Wash and circulate while monitoring wellbore pressure. If you tag solids or obstructions above the perforations, pick up and check the tubing string weight. Check the circulation to ensure good continuous fluid returns.

Slack off slowly and tag solids lightly to find out if the tag-up spot is a steel object, sand, or scale elements. Do not apply too much force on the tagged-up spot. If the object cannot be washed or drilled away in a few short minutes, you may consult with the company man. (Note: Observe limited allowable push or pull onto the coiled tubing string.)

Avoid too much weigh on the bottom-hole assembly, causing buckling on the coiled tubing (avoid forcing the bit to sidetrack). Drill out and break down solids. Circulate sand and solids out of the wellbore. Catch a sample of the material coming out of the well (check for steel shaving, rocks, shale, or formation sand in returns). Steel shavings will indicate that you are cutting on a tight spot in the tubing/casing and/or ***a*** lost steel object in the well (you may install a magnet at the fluid returns to catch steel samples).

Drilling or milling on any tight spots in the casing string may cause the mill to sidetrack and take off into the annulus or formation. You may get the coiled tubing stuck and/or lose the wellbore (Glenn vs. the Pillar 5 saltwater disposable well).

Do not push or slack off too much weight on the coiled tubing. Maximum weight must be approved by the coiled tubing manager to avoid downhole problems. Continue washing sand, mud, and solids through the open perforation. Pick up and check the circulation. Circulate bottoms up if possible before washing through the open perforations.

Types of wellbore pressures:

- hydrostatic pressure
- formation pressure
- differential pressure
- trapped pressure
- friction pressure
- bottom hole pressure

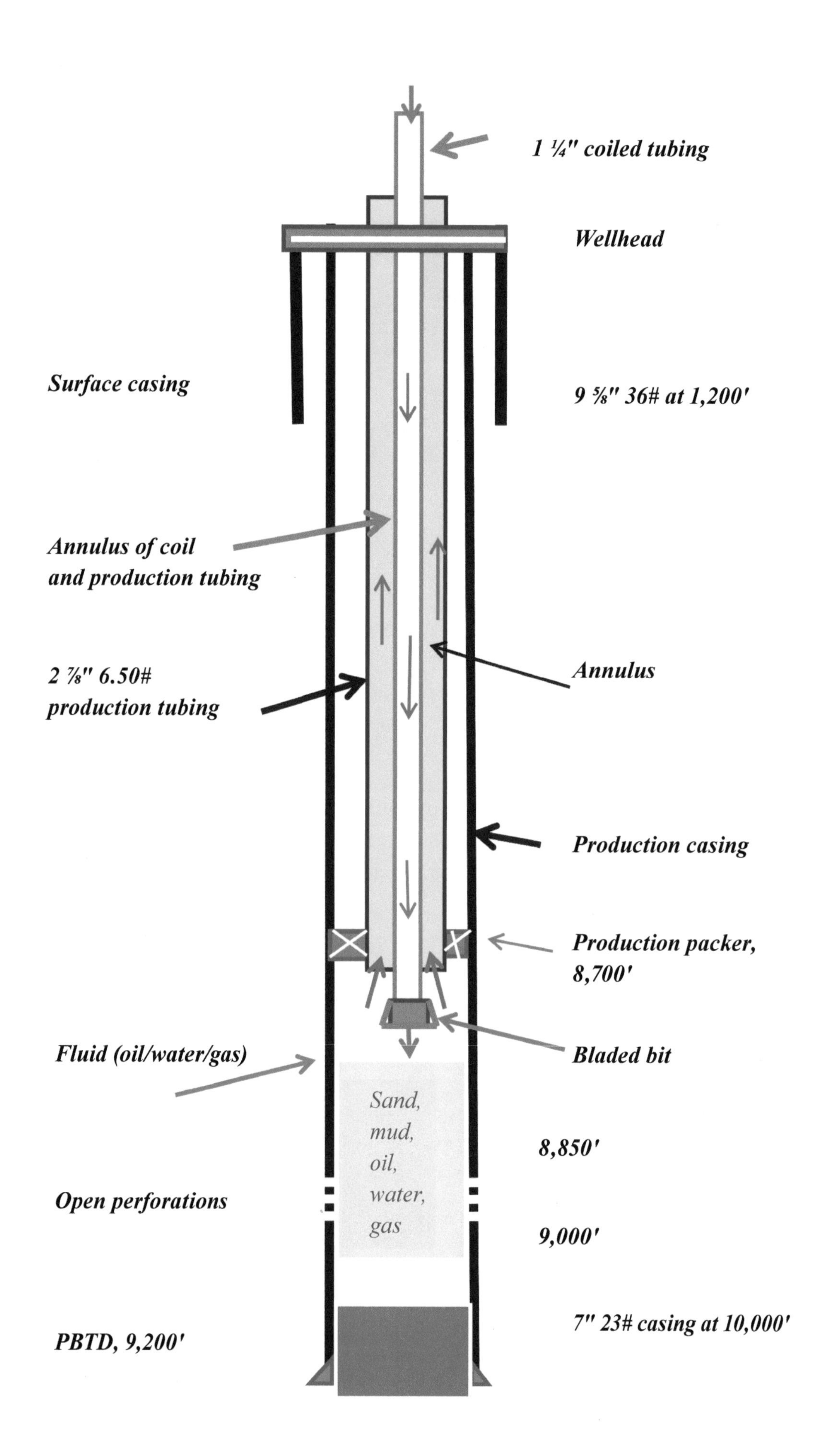
1 ¼" coiled tubing
Wellhead
Surface casing
9 ⅝" 36# at 1,200'
Annulus of coil
and production tubing
2 ⅞" 6.50#
production tubing
Annulus
Production casing
Production packer,
8,700'
Fluid (oil/water/gas)
Bladed bit
Sand,
mud,
oil,
water,
gas
8,850'
Open perforations
9,000'
PBTD, 9,200'
7" 23# casing at 10,000'

Some reservoirs will have very low bottom-hole pressure. Hydrostatic fluid pressure in the well will cause the loss of circulation and may bridge off sand/solids around the coiled tubing and get the coiled tubing stuck. Some reservoirs may have a significant high pore pressure and may kick on you when washing solids just above the perforations.

Always wash and circulate sand and solids slowly, not faster than 10′ per minute at the bottom to allow the solids to break apart with the fluid. The wellbore gas pressure will assist you in the circulation and will lift solids out of the wellbore, similar to nitrogen gas.

Continue washing to below the open perforation while checking the fluid and watching the flow for abnormal reservoir influxes. If you are using nitrified fluid to lift sand, oil, gas, and water, check your circulation more closely for abnormal gas and liquid flow into the rig tank (nitrogen will reduce hydrostatic pressure and make the well flow unexpectedly).

Once the wellbore is circulated and clean, cut the nitrogen off while circulating with the liquid at higher rates. Continue to pump liquid and circulate the nitrogen out of the well bore to keep pressure under control.

Flowing Oil Wells and Gas Lift Wells

Monitor the wellbore pressure and start out of the hole with the coiled tubing and the bottom-hole assembly. In a low-pressure oil and gas wellbore cleaning operation, you may be asked to jet the liquid with nitrogen while pulling out of the well, allowing the well fluids to flow into the available fluid tank or simply into the production line.

Jetting out the wellbore liquid will reduce hydrostatic pressure and enables the reservoir fluid to flow naturally!

Trip out the hole slowly above the crown valve (swab valve). Close shut the crown valve only and allow the well to flow through the production line, or you may continue flowing the well to unload into the rig tank, as instructed by the person in charge of operations.

Displace and unload the kill fluid or wash water out of the coiled tubing with nitrogen gas into the rig tank. Rig down the coiled tubing tools and equipment. Clean up the location and put the well's monitoring equipment just the way it was. Turn the well over to production and move out of the field safely.

If the purpose of the coiled tubing intervention is to clean up a gas well and/or oil well, follow safety instructions from the company man on the location. On a low-pressure gas well, generally, the well bore will be dewatered by using nitrogen after cleaning. The well fluid will be displaced by nitrogen while coming out of the well. On oil-producing wells, the wellbore will be cleaned up and displaced with fresh fluid. The coiled tubing will be pulled out of the well safely, and the well will be lifted after.

Well control in case of emergency:

- Inform the company man and your supervisor immediately.

- Shut the well in and check for any possible leaks.
- Wait for instructions while filling up and checking prerecorded wellbore data in the kill control kick sheet.

There are several choice methods of killing a well based on the policy and procedure of the oil and gas well operator:

- The drillers method
- The wait and weight method

These well control methods involve keeping the bottom-hole pressure constant while allowing the influx to rise to the surface.

- Bull-heading drives the influx back into the formation. Bull-heading crude oil into the formation often becomes difficult (depending upon the oil viscosity).

Bull-heading of kill fluid in coiled tubing operation is conducted during emergency occassions only"

- stuck coiled tubing off the bottom
- parted coiled tubing string at the surface
- collapsed coiled tubing at the surface

Do not bull-head kill fluid without having a careful plan in place (bull-heading kill fluid requires experience and knowledge with sound planning).

HP (hydrostatic pressure) per barrel of kill fluid:

$$HP = \frac{53.45 \times MW \text{ (psi/bbl)}}{D_h^{2} - D_p^{2}}$$

***

Elements Coming Out of Oil and Gas Formation Reservoirs

a. Crude oil (rock oil) is an organic compound that appears in different colors and textures (brown, black, green, and red with different gravities).

b. Saltwater is the major unwanted element that comes out of a wellbore. Saltwater will appear in different densities. Some wells produce considerable amount of saturated salt buildup with oil production and small amounts of saltwater. The production of salt will form hard salt bridges for several hundred feet. Fresh water treatments or hot water applications will prevent the salt deposits.

c. Natural gas is a naturally occurring element primarily driven from methane (CH4) and other combined gases such as ethane, propane, butane, H2S, helium, argon, and carbon dioxide.

d. Hydrates are basically the products of wet gas forming into hard solid compounds that may completely plug off the flow line, tubing, chokes, and checks and/or well control equipment. Hydrates may become major problems if they are not removed quickly. Hot water and/or other types of chemical inhibitors can remove the hydration deposits.

e. Paraffin is basically an oil byproduct (you can melt it and sell it with the oil). Paraffin buildup is due to the change in physical and chemical equilibrium of crude oil coming out of a well. Primarily, paraffin is in a solution in the reservoir because of higher temperatures. As the crude oil flows up the wellbore, the paraffin will come out of the solution because of temperature change and will be deposited inside the production tubing and the casing string.

   Paraffin is an aggravating problem in oil production during cold climate. Some wells may have severe paraffin problems several hundreds of feet from the surface down the well and may completely plug off the production and flow lines. Hot oil and/or stems will remove paraffin buildups. Paraffin will melt at a temperature of 150–200 degrees.

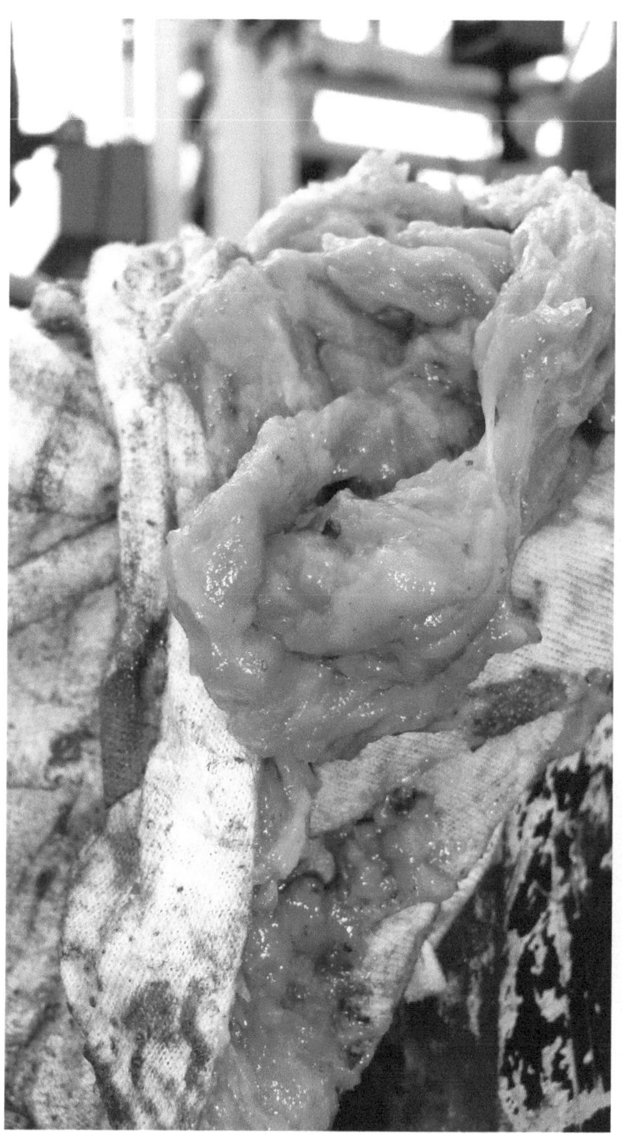

f. Formation sand is one of the major problems in the Gulf Coast oil and gas well. Formation sand is an unwanted material produced with oil, water, and gas. High-producing artificial lifts are the main cause of high-rate formation sand in the wellbore. Gravel packing is the best prevention method to reduce wellbore cleaning and avoid casing problems.

g. Sand and scale deposits are one of the major problems in deep oil and gas wells. Calcium carbonate, calcium sulfate, and barium sulfate are tough to deal with. Some sulfate deposits are very difficult to stop without the necessary chemical applications. Acid chemicals will not penetrate barium sulfate and must be cut or drilled out using coiled tubing operations.

Advantages and Limitations of Coiled Tubing Applications in Oil and Gas Operations

The advantages of coiled tubing services are many:

- The unique and key advantage of coiled tubing is the ability of continuous fluid circulation while going into a well and coming out of a well.

- It entails quick rigging up and rigging down compared to conventional workover service rigs and drilling rigs.
- Circulating while going into the well and circulating while coming out of the hole is a significant safety alternative advantage over the other choices.
- Using the continuous coiled pipe with no tubing collar connection will save time and is a cost-saving factor, safer and faster than the conventional workover rig operation.
- Intervention solutions can be achieved without killing a well or pulling the production tubing string out of a well using a coiled tubing string. You will save time and money compared to conventional workover pulling units that require pulling tubing joints, hydro-testing the tubing going back into the well, and cost of packer repairs and rented tools and equipment.
- Using coiled tubing will reduce well control risks of minor kicks, blowouts, and pollution (constant well control measures going into and coming out of a well).
- The application of coiled tubing in the open hole gravel packing is safer, quicker, and more efficient (if it is done correctly).
- Coiled tubing is used to continuously wash and circulate a wellbore while going into a well and to continuously circulate or displace fluid out of a well while tripping out of a well.
- Coiled tubing is used to drill, wash, and circulate out drilling mud, sand, and abrasive solids out of the wellbore using plain water and/or nitrified water. Most coiled tubing operations used for washing, saltwater disposal, or injection well may require only eight hours for a job well done. It would require four to five days for a workover rig operation to complete the same work task by pulling and running the tubing string (significant cost saving).
- Coiled tubing is used for jetting and de-liquidizing the gas wells and/or oil wells using nitrogen to save the cost of pulling tubing and/or avoid swabbing operations (reduce hydrostatic significantly to bring the well in).
- The entire deep wellbore can be circulated dry using a foam/nitrogen application.
- Note: Never use dry air to wash or clean up oil and gas wells. Dry air consists of oxygen and several gases that may not be compatible to oil and gas wellbore cleaning operations. Introducing dry air may cause chemical reactions and create unbelievable explosions (you do not want to witness that).
- Coiled tubing may be used to drill or mill out hard scale deposits.
- Coiled tubing is used to operate (open or close) sliding sleeves and zone isolation techniques. Cleaning a horizontal wellbore can be completed successfully using coiled tubing. A coiled tubing string may be more flexible, going through deviated holes with superior performance, than tubing joints in conventional workover rig operations.
- Coiled tubing is used to spot cement plugs or acidizing a well. The operator must use adequate corrosion inhibitors to protect the pipe.
- Note: High concentration acid such as HF or HCL will shorten the useful lift of coiled tubing.
- Coiled tubing is used to perforate wells using tubing-conveyed perforating guns (TCP).
- Coiled tubing is used in various types of wellbore treatments to pump chemical products or spot cement above a plug.
- The application of coiled tubing using nitrified fluid will avoid the loss of water in a sensitive reservoir formation (some reservoirs may be highly sensitive to water).

Coiled tubing may be used as capillary string and/or production strings in low-pressure oil and or gas wells. The coiled tubing velocity string is an ongoing project in low-volume, low reservoir pressure wells in oil and gas fields.

- A coiled tubing string is used to drill horizontal wells with great success to save time and reduce costs. It is an effective and efficient practice.
- The application of coiled tubing to wash, drill, circulate, and clean horizontal wells after fracturing operation is an efficient and cost-effective practice.
- Coiled tubing may be used for pipeline cleanup, offshore and onshore.
- Use coiled tubing in under balance completion.

***

Disadvantages and Limitations of Coiled Tubing Applications

What can go wrong when you using coiled tubing?
Nothing can go wrong if it is done right!

Using inferior-quality equipment and inexperienced coiled tubing personnel can complicate the operation (anything may go wrong).

1. Using coiled tubing in major well control is a concern.
    a. Smaller sized coiled tubing has a thin wall and is subject to becoming parted, bursting, or having holes because of repeated and continuous pipe motions under tremendous stress and strain forces in and out over the gooseneck (major well control problem).
2. H2S and high concentrations of HCL or HF acid and/or acidic chemicals will cause deterioration in the coiled string more quickly than you may think.
3. Coiled tubing cannot rotate by itself from the surface ground. Working the coiled pipe up and down is the only viable movement. Hydraulic motors and a bottom-hole assembly can be connected to the coiled tubing and used to conduct the drilling or milling operation successfully. It is often difficult to monitor the performance of the bottom-hole assembly.
4. Coiled tubing is small, thin, and fragile; it is not stiff enough to perform certain tasks. Pipe may twist and turn in different shape and forms while going down the tubing and casing causing kinks and deformation
5. (the coiled pipe is viewed as the weakest of the components on the coiled unit).
6. Coiled tubing is subject to various surface and subsurface acting forces during the intervention operation, including the following:

- ➢ Pushing force
- ➢ Pulling force
- ➢ Bending force
- ➢ Straightening force
- ➢ Slide friction force
- ➢ Slips drag force
- ➢ Injector force
- ➢ Wellbore fluid force
- ➢ Buoyant force
- ➢ Snubbing force

Force on the coiled tubing is either pushing in the hole or pulling out of the wellbore with considerable stress fatigue.

- The reel force is the force that takes place over the drum while spooling over or off the reel. Constant bending and straightening forces will deform the coiled tubing and will create an egg shape. You may not be able to run a cutter through an oval-shaped pipe.
- The injector force is the force applied to the coiled tubing while curving over the gooseneck assembly.
- The third force is the straightening after passing over the guide arch and into the injector (repetitive bending and straightening forces going in and coming out).
- The snubbing force is the necessary force of pushing the coiled tubing through the injector head and down the well.
- The buckling force is the major ongoing action taking place when snubbing the pipe through and/or tagging and slacking off on the pipe, causing different shapes of looping and bending inside the production tubing or larger casing string.
- Internal and external friction forces include fluid friction from the inside of the pipe and sliding, rolling, and solids friction from the outside of the pipe.
- Differential pressure is the pressure difference between the inside of the coiled tubing and the pressure outside of the coiled tubing (the annular space).
- High abnormal differential pressure may cause the coiled tubing to burst or collapse.

1. Coiled tubing is subject to burst under higher pressure because of wall loss or wall thickness reduction and may become parted.
2. Note: Coiled tubing may part and unwrap because of potential tension energy from being stored over the reel.
3. Stored tension energy at the spool above the ground may be subject to unwind on the reel and create accidents or damage. Parted tubing at the reel because of pipe defects and/or the operator's error is dangerous.
4. Coiled tubing is under constant high injection pressure because of the inside diameter of the coiled tubing and the restrictions of the bottom-hole assembly (high pumping rates will cause higher friction pressure and internal wear).
5. Coiled tubing is subject to become parted because of fatigue from constant cycling motions over the reel, the gooseneck, and the stripper head. Coiled tubing fatigue is due to repeated motions of rubbing and sliding over the reel, the gooseneck, and the injectors system. The tubing may become oval shaped.
6. Coiled tubing is a seam-type pipe (ERW). Manufacturing defects, age, wear, and misusing the pipe can lead to failure in any well.
7. The pump rate is low, and the pressure is too high because of the long length of the continuous small pipe with high friction.
8. The application of coiled tubing in internally coated tubulars is a concern. Coiled tubing is subject to cause damaging friction to internally coated production tubulars because of twists, turns, drag, slide, and buckling effects.
9. In the doglegs and crocked/deviated holes, the coiled tubing may cause great damage to an internally plastic-coated tubing string.
10. For your information, there is no such thing as a straight hole drilling in the oil field these days; it is just controlled drilling (often missing the target, and we call it dry hole) three or

four degrees from the vertical line to be a straight hole. The only straight hole to the target I know is the hole drilled (tunnel) from the country of Mexico to a house in Texas.

11. Light coiled tubing and wireline is subject to be forced and lifted out of the well because of high wellbore pressure. This may occur when a small length of empty light pipe enters a high-pressure wellbore condition. The pressure will force the coiled pipe, like a piece of straw, out of the injector head toward the coiled tubing reel, creating large pipe loops over the injector all of a sudden.

    Similar scenarios occur when running wireline tools into the well under high wellbore pressure. As soon as you open the master valve, the high pressure will force the tools and wireline into the lubricator and cause a huge nest inside the lubricator (blowouts and fires are possible). Pulling a packer out of the casing string under high reservoir pressure will force the tubing to run away toward the crown. (You may be lucky if you manage to shut the well in.)

12. Early detection and quick reactions are great prevention methods to avoid accidents.
13. Most coiled tubing failures may take place above the ground (the space between the Christmas tree and the coiled tubing service reel).
14. The major reasons for most coiled tubing failures above the ground are the bending and snubbing forces that take place between the reel and the injector head.
    - Parted coiled tubing between the gooseneck arch and the service reel
    - Parted coiled tubing between the stripper head and the injector
    - Buckled coiled tubing between the stripper head and the injector
    - Leak in coiled tubing between the tubing gooseneck and the real
    - Leak in coiled tubing between the tubing arch and the stripper assembly
    - Patching and welding problems
    - High circulation pressure due to pipe inside diameter (ID)
    - Reduction of wall thickness due to tensile strength
    - Under a constant fatigue during operation

    Flattened Pipe due to differential pressure:

Coiled tubing is not as straight as you think while going down the hole.

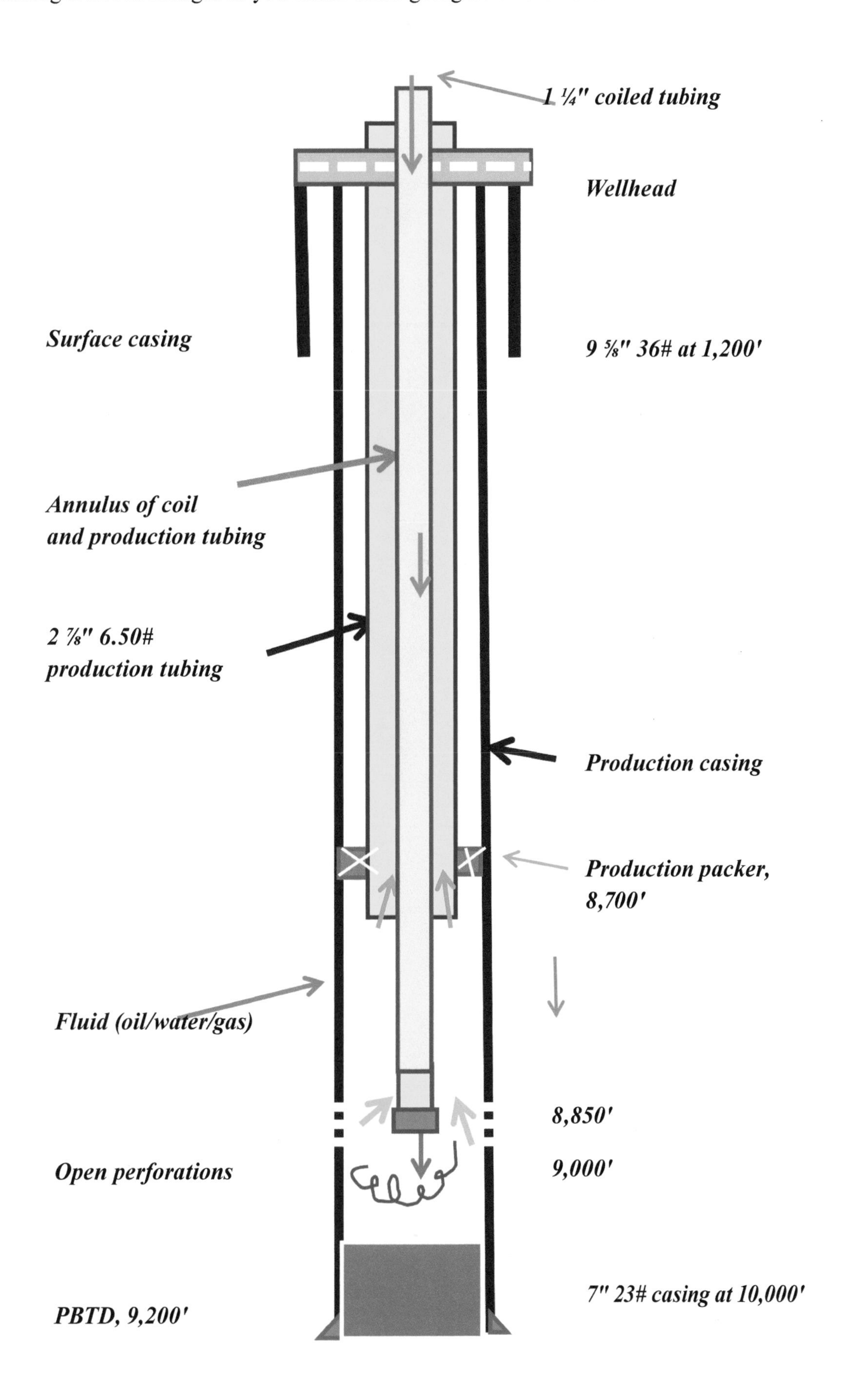

Actual Downhole View of Coiled Tubing Going Down the Hole

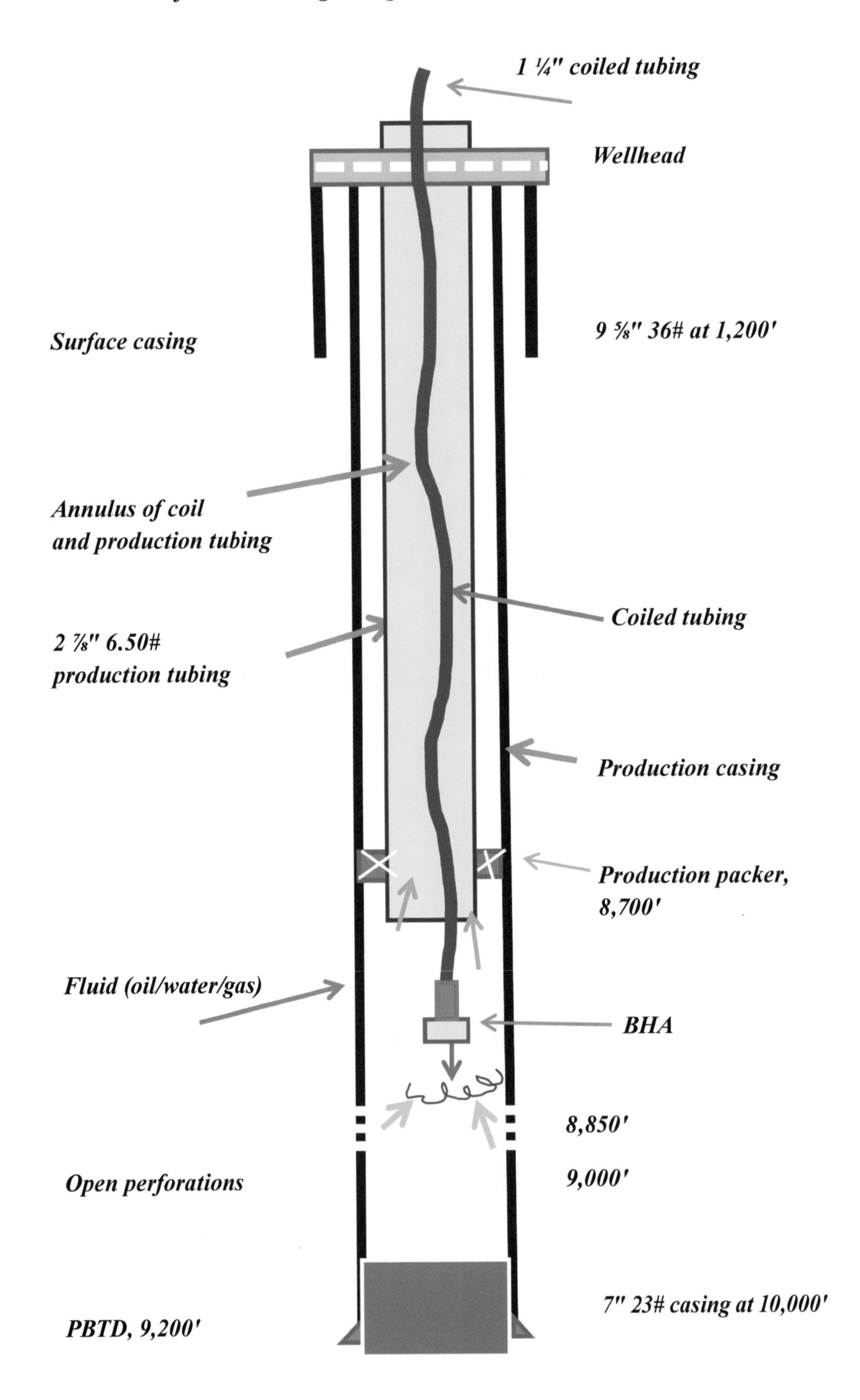

CPSIA information can be obtained
at www.ICGtesting.com
Printed in the USA
BVHW021951200120
570016BV00014B/68

9781796071863